Firing kilns

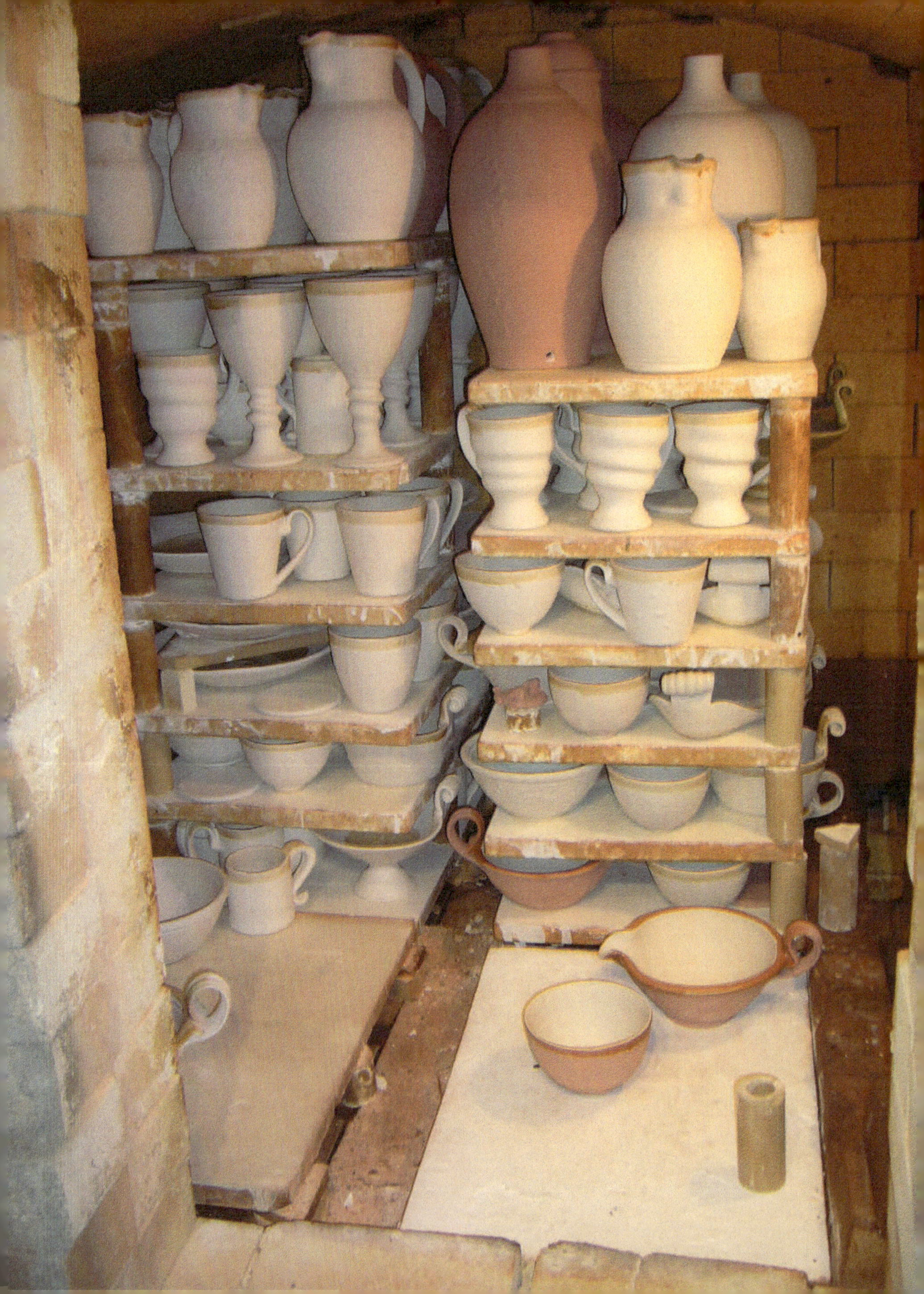

Firing kilns

Benedict Brierley

B L O O M S B U R Y
LONDON • NEW DELHI • NEW YORK • SYDNEY

Bloomsbury Visual Arts
an imprint of Bloomsbury Publishing PLC

50 Bedford Square
London
WC1B 3DP
UK

1385 Broadway
New York
NY 10018
USA

www.bloomsbury.com

A catalogue record for this book is available from the British Library and the U.S. Library of Congress

ISBN: 978-1-4081-8524-7

Designed and typeset by Susan McIntyre
Printed and bound in Great Britain

MIX
Paper from responsible sources
FSC® C013604

FRONTISPIECE: John Calver's oil kiln half-packed using a three-prop stacking method. *Photo: John Calver*

Contents

Acknowledgements

I would like to thank all the artists and makers who have contributed so generously and enthusiastically to this book: Alistair Young, Andrew Niblett, Antje Earnestus, Christine Gittins, Dylan Bowen, Eddie Curtis, Froancoise Dufayard, Gabriel Koch, Helen Marton, James Hake, Jim Gladwin, David Scott, John Calver, John Wheeldon, Lone Borgen, Marcio Mattos, Matthew Blakely, Michael Flynn, Monika Patuszynska, Nigel Lambert, Ruth King, Sarah Villeneau, Oscar Copping, Thomas Aitken, Kate Hyde and Stephen Parry for their time, wealth of knowledge and images of their kilns and work.

I would like to thank Debbie Claxton for her help with the index, and Loughborough University School of the Arts for their support of this project and for use of their facilities. Finally I'd like to thank Bloomsbury for giving me the opportunity to realise this book.

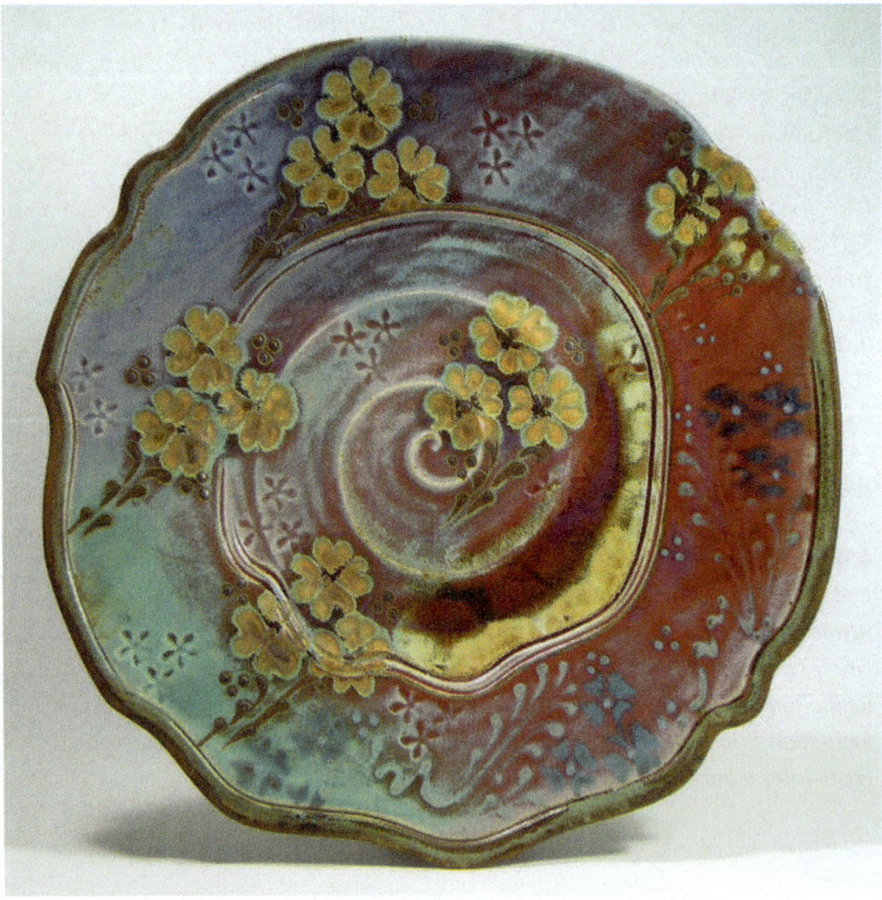

Platter, John Calver.
Diameter: 40 cm (15¾ in).
Photo: John Calver

Introduction

It is believed that clay was first fired and transformed by heat into ceramic around the late Stone Age period, Upper Palaeolithic 30,000–10,000 years ago. Although the details of the events which instigated this discovery will forever be the subject of speculation, it is thought to have coincided with the development of agriculture and the departure from a hunter-gatherer existence in favour of more settled communities. It is reasonable to surmise, however, that the discovery of this material and its ability to change into a durable and functional substance would have been a huge technological step forward in allowing communities to produce permanent high-status objects and vessels for ceremonial and domestic purposes.

The earliest evidence of this transformation of the material appear as effigies of human and animal forms thought to have been ritual objects and talismans passed down through generations to protect the owners through childbirth. Evidence of utilitarian vessels formed and fired from clay, appear in Southern China and date back around 20,000 years. Although these have been the oldest verifiable evidence of early use of ceramic, most cultures have their own history of the use of clay as a material for the creation of objects. The range of uses found is huge, from fundamental building materials such as brick and tile to exquisite vessels and sculptures.

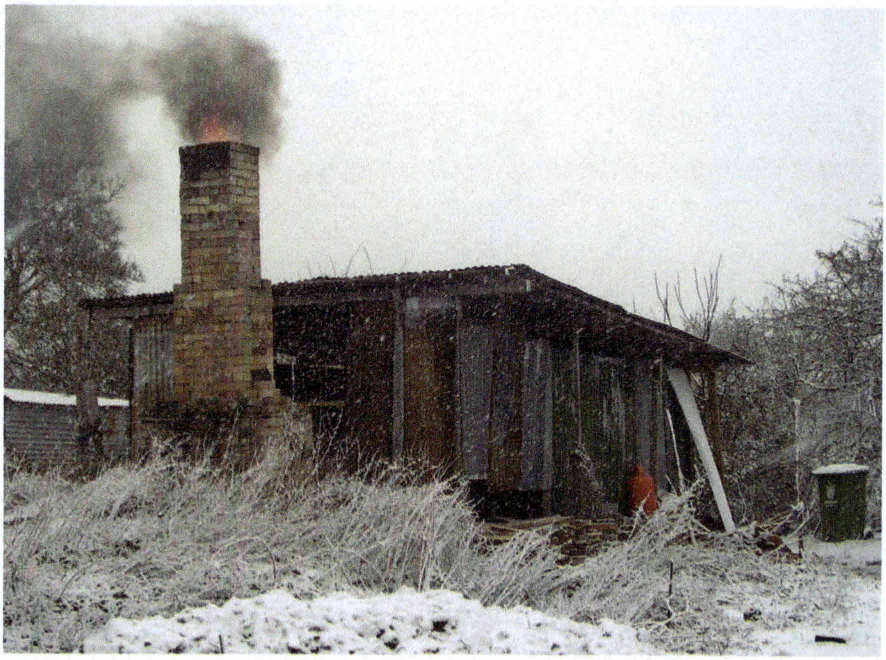

Wysing Arts Centre UK. Anagama kiln firing. *Photo: Author*

The material which we use as ceramic makers today has therefore been around for a very long time and the technologies for working it and rendering it permanent began early in the emergence of modern humans. The structures that have over thousands of years been developed to alter the state of clay would have arguably begun as small pits in the ground filled with artefacts or pots and then covered with brush wood and ignited. Over centuries of successful and unsuccessful firings these simple kilns evolved, arguably driven by local economic and domestic factors into what we now recognise as ceramic kilns. Essentially a kiln is a structure which contains heat efficiently enough to generate an internal ambient temperature capable of activating the chemical changes which occur in clay when it transforms to ceramic.

Over the last 50 years or so many older kiln technologies have been revisited by contemporary makers and this exploration has expanded our knowledge and understanding of how these structures were used by the cultures which developed them and how the visual and structural qualities of work which is produced using these technologies varies in its visual and aesthetic qualities. It has also expanded hugely our appreciation of the many and varied ceramic surfaces which can be generated by these technologies and how these surfaces and firing techniques may be applied to the work of makers working today.

This book will attempt to demystify the process of firing clay and how kilns may be used to achieve a multitude of diverse clay and glaze surfaces, it will examine some common firing faults and how these may be remedied as well as exploring the many branches of the kiln's evolutionary tree and how these variants are used by contemporary practitioners to produce their work in the 21st century.

RIGHT: *Raku-fired lustred vessel,* John Wheeldon. Lustre fired to 650°C (1202°F) in a propane raku kiln, height: 18 cm (7 in). *Photo: John Wheeldon*

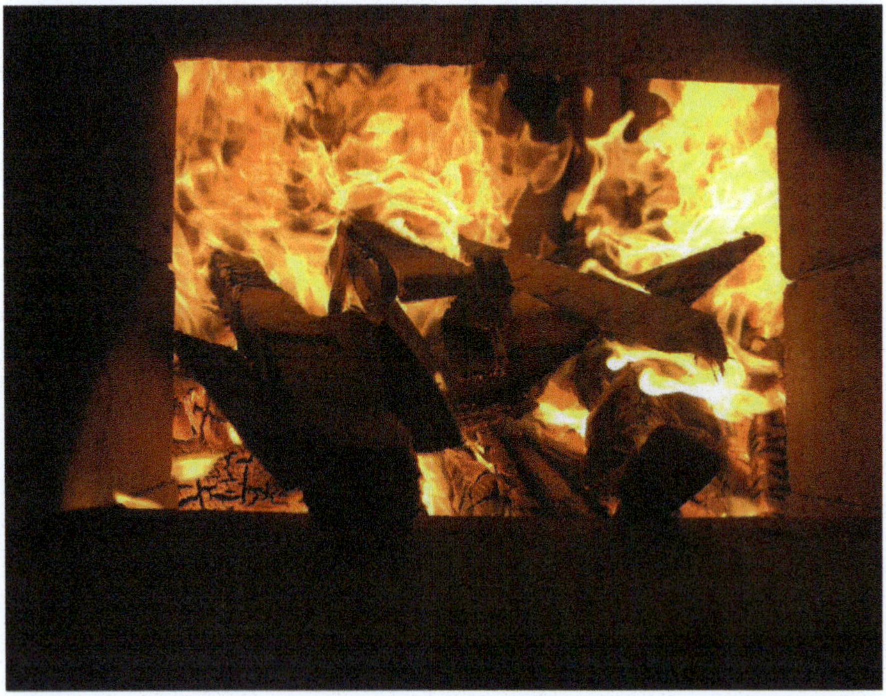

LEFT: The early stages of wood firing in a crossdraft kiln. Building a bed of embers at the front of the kiln with a fairly low ambient temperature further back. *Photo: Author*

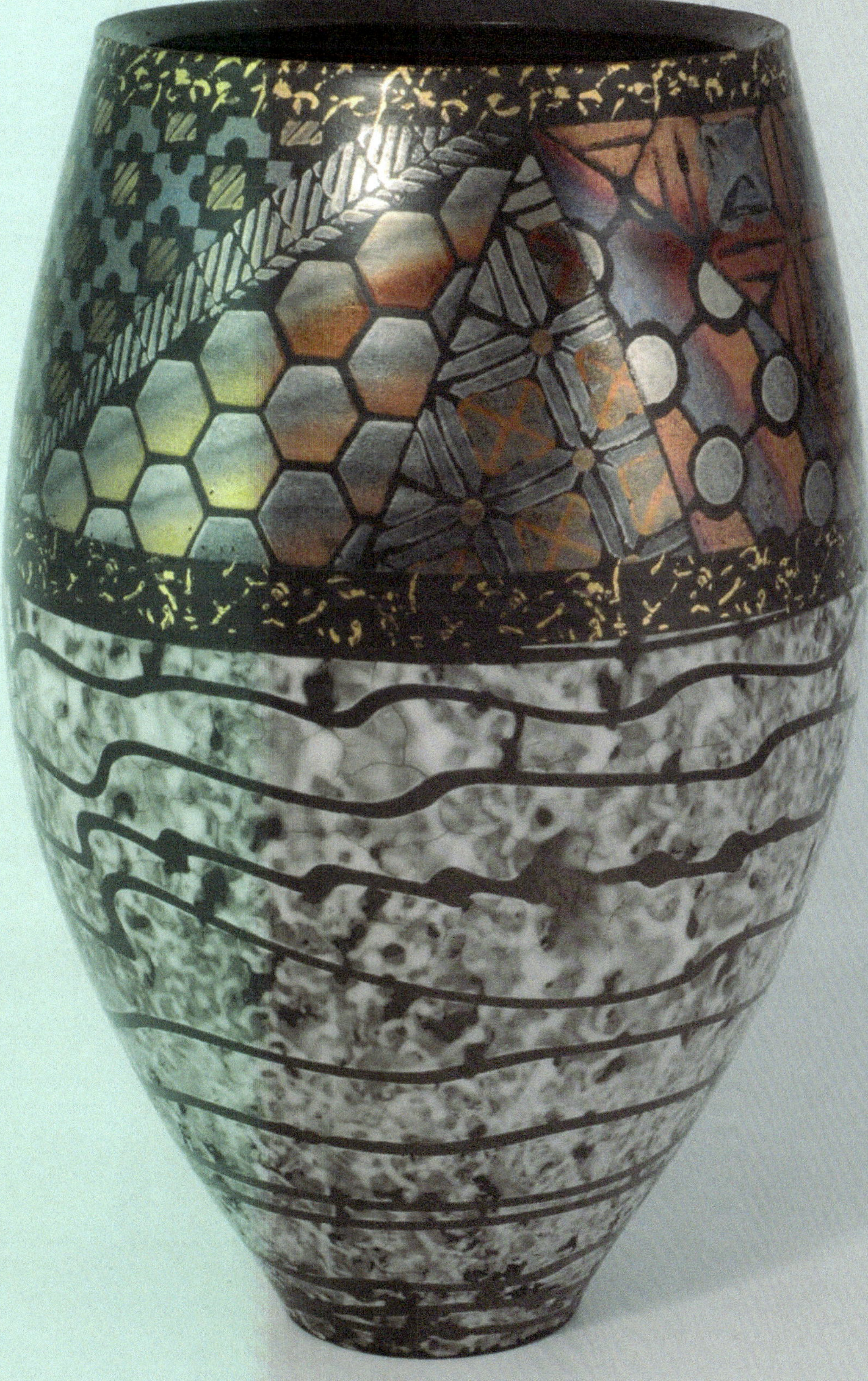

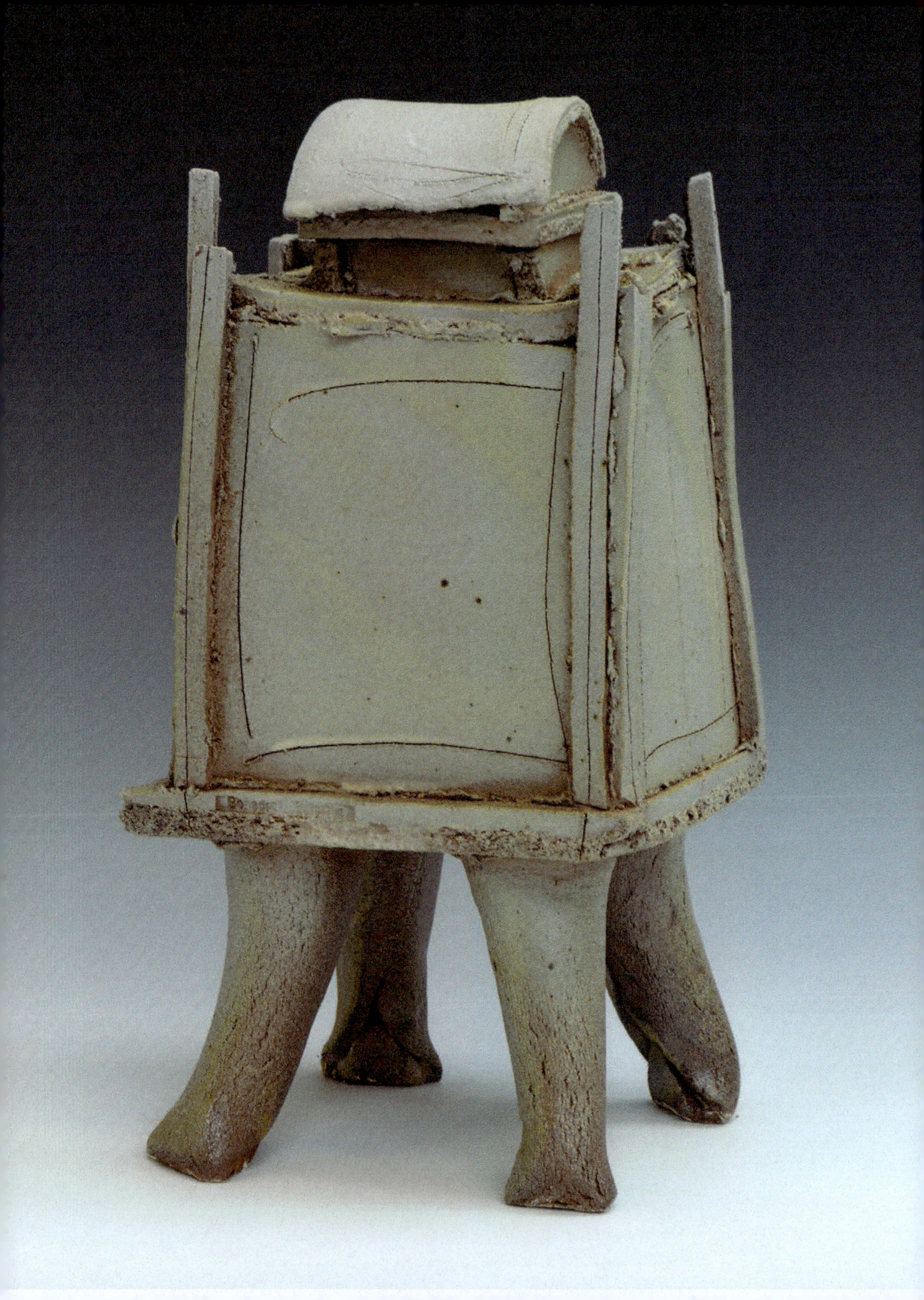

1

Firing principles

WHAT IS CLAY?

Clay is an aluminosilicate plastic material derived from the decomposition of predominantly silica bearing rocks. This decomposition is propagated by the elements: heating, cooling, weathering and the action of weak acids which together, over millions of years, break down the parent rock to create deposits of primary clays known as kaolinite (china clay). If these deposits are subsequently moved by the action of water in the form of streams and rivers, the resulting deposits are known as secondary clays or, as they are more commonly termed, ball clays (a term which describes the shape the clay was made into when it was mined). These clays may be deposited many miles from their source and contain more impurities than the primary clays. As well as picking up impurities such as iron, titanium and magnesium from the geology beneath the water courses, the secondary clays also collect organic matter and, due to the action and abrasion of waterborne sediments, are ground into finer particle sizes. The relatively finer particle sizes of secondary clays are what give them their high plastic qualities. It is possible for secondary and primary clays to be used straight from the ground, although they may contain elements which make them hard to work or that will have a negative effect in the firing process. Contaminants such as mica or high calcium deposits as a result of tiny seashell remains may limit the plasticity or fired properties of the clays.

LEFT: *Lidded Box,* Benedict Brierley. Slab-built and wood-fired for 3 days in a crossdraft kiln. Height: 45 cm (17¾ in). *Photo: Author*

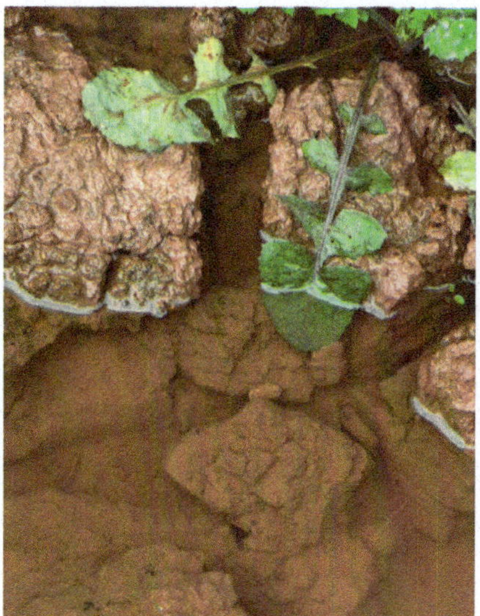

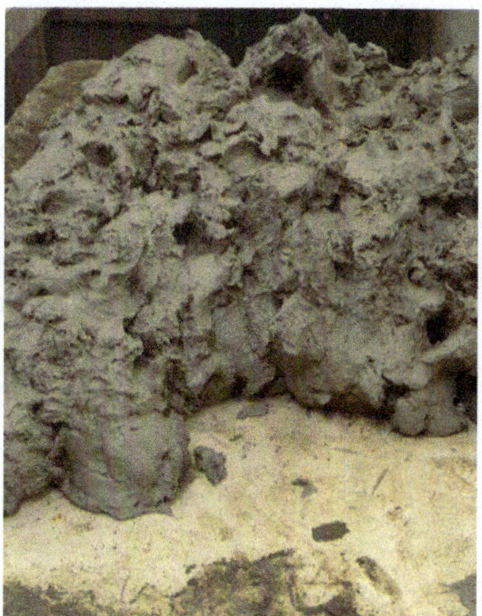

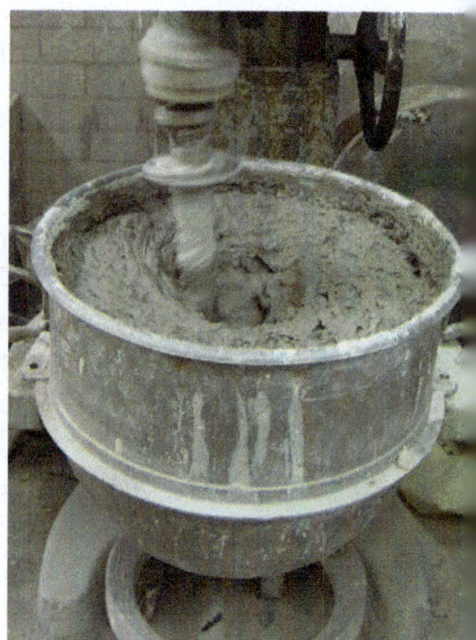

BELOW, LEFT TO RIGHT:
Raw red clay.

Slaked clay drying on a plaster of Paris slab: in preparation for pugging.

Bespoke clay being mixed in a dough mixer.

Photos: Author

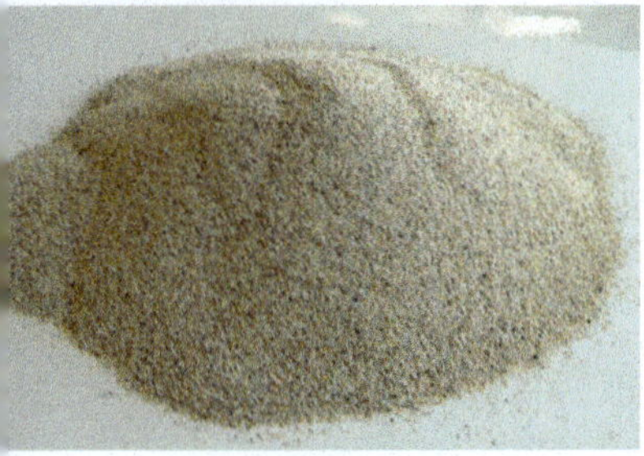
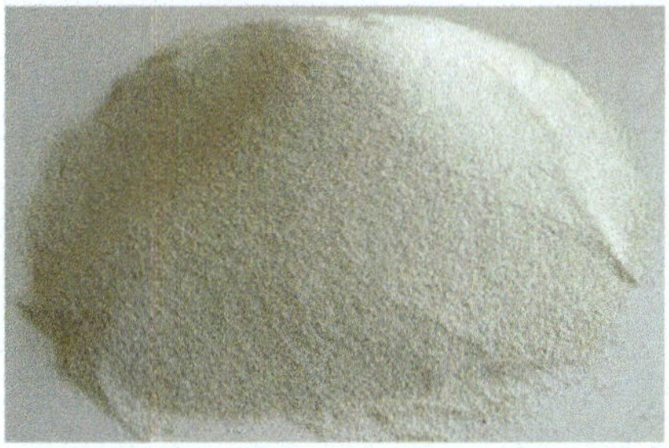
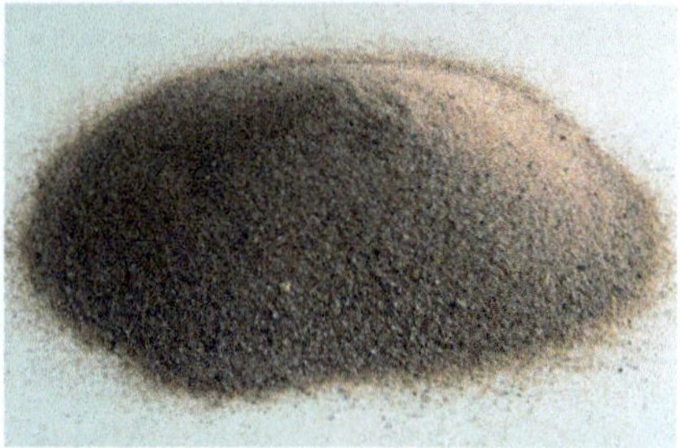

Most clays commercially available today, either in dry or plastic forms, are blends of primary and secondary clays, which have been processed to remove any contaminants that may decrease the workability and fired qualities of the body. They may also have additions of silica and grog, the grog usually being pre-fired ground clay or sand included to open up the body for textural qualities and to increase resistance to thermal shock. Porcelain, for example, contains 50–60% china clay and it is this high degree of pure primary clay that is so beneficial in the development of a hard, white body. A consequence of a higher percentage of primary clay containing larger particles, however, is to render the material less plastic than many other blends of clay. To counteract low plasticity, some porcelains will have a small amount of low-iron secondary clay or bentonite (an impure clay of very high plasticity), added to improve the workability of the body, and also a little whiting (calcium carbonate), which help to bleach out any discolouration from iron in the added clays. Red terracotta on the other hand, which is made from pure secondary clay, has particle sizes that are much finer and therefore the material as a whole is far more plastic and the red colour indicative of the amount of iron that the clay collected on its travels from the source. Many types of red clay can be used straight from the ground.

ABOVE, LEFT TO RIGHT:

80–30 mesh fireclay grog for adding to clay bodies. The grog effectively opens up the body adding strength, reducing shrinkage and promoting even drying.

80–30 molochite grog, a pure white grog made from calcined china clay.

Fine silica sand grog. Sand may also be used when placing wide-based objects in the kiln allowing the pieces to move freely as they shrink.

Photos: Author

EARTHENWARE AND STONEWARE

Before we proceed, it may be prudent to explain a few definitions relevant to clays and firings and how they are applied. Two terms commonly referred to are stoneware and earthenware. These descriptions can be applied to both clay bodies and firings. They will have slightly different connotations depending on which is being discussed. Clays are usually described as either earthenware or stoneware. An earthenware clay body will be said to fire between 1000°C and 1160°C (1832–2120°F). If it is a true earthenware body, firing higher than this top temperature will cause the body to become soft, eventually becoming a glaze. A stoneware clay will generally have a firing range from 1000°C to 1300°C (1832–2372°F) or even higher. Eventually, even with a stoneware body, if fired high enough it will break down and become glaze-like. So an earthenware body generally has an upper limit of mid-1100s°C (2100°F), a stoneware clay mid-1300s°C (2500°F).

When the terms earthenware and stoneware are applied to firing temperatures, they take on a slightly different meaning. An earthenware firing could be said to be

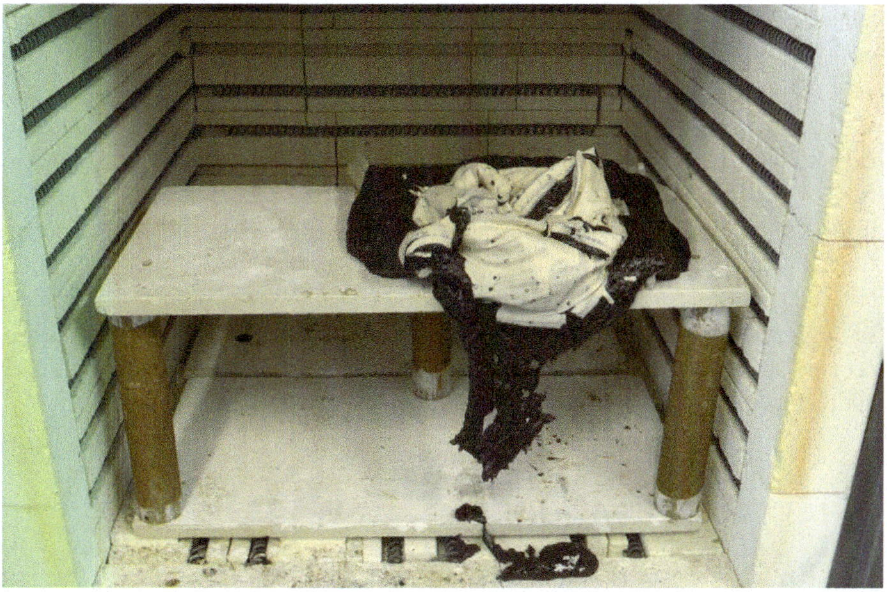

Jar thrown from Charnwood brick clay with white slip (50/50 by weight: ball clay and china clay). Over-fired to 1260°C (2300°F) in electric kiln. *Photo: Author*

carried out between about 1020°C and 1180°C (1868–2156°F), any clay can be used – the term merely refers to the temperature to which the firing is being carried out or the temperature that an applied glaze melts at. A stoneware firing temperature similarly refers to a temperature range from 1180°C up to around 1300°C (2156–2372°F), and for this type of firing, stoneware clay should be used because, as stated earlier, an earthenware clay will begin to melt above 1180°C (2156°F) (and many a lot lower). Clearly, if earthenware clay is placed into a stoneware firing, the chances are that the result will at best be a very deformed pot.

PREPARING CLAY FOR FIRING

The first stage in preparing a clay object for firing is to dry the clay. This may be achieved by leaving the work in a warm environment with good ventilation for several days. As the material dries there is a degree of shrinkage, which will vary depending on the type of clay being used, with finer clays having a higher shrinkage rate than courser clays. Shrinkage occurs due to the water held between the crystalline clay platelet structure evaporating; hence the plates move closer together. There will usually be around 5–10% reduction in size overall as the water is lost. The water that lies between the clay platelets, while present, acts as a lubricant, allowing the plates to move against each other and giving clay its plastic malleable qualities. As the clay dries, it becomes stiffer and changes state from plastic and soft to leatherhard and finally to dry.

Once the material has lost all its water and changed colour, it can be described as 'green ware', and is ready to be fired. For many of the earlier centuries of ceramic production, firing processes simply allowed for a more durable vessel or object to be made out of an abundant natural material, which was fairly simple to prepare and work. Even in these early objects the decorative potential of clay was being explored, with some of the very earliest vessels having rope decoration impressed into the clay surface. Many early hand-built vessels also used the making marks that were unique to each maker to decorative ends. As ceramic understanding evolved, with better control over firings and materials, the potential of glazes and the additions of colouring oxides to these glazes not only increased decorative potential, but created even more functional objects. Early on in ceramic history the technologies did not exist to fire clays up to their vitrification point (the point at which the clays become waterproof), so the development of glaze and the ability to fire it successfully, rendering the porous ceramic surface waterproof at the lower temperatures that were achievable, was another major advancement.

WHAT FIRING CLAY DOES

Once clay is placed in a kiln and heated, it is subject to several physical and chemical changes. The first is that all remaining water is driven off; it is for this reason that all work must be thoroughly dried before being fired. A useful way of testing whether a piece is dry enough to be fired is to warm the piece up in an airing cupboard or warm oven, or even sunshine, and then place the object against a cool piece of glass such as an exterior window. If any condensation is left on the glass the piece should be dried for longer, and periodically tested until there is no condensation.

Once the temperature rises above 100°C (212°F), if any water is left in the clay it will turn to steam and expand massively (expansion 1:1600–1700). Therefore, if the climb or ramp rate of the firing is too fast in the early stages of a biscuit firing the work will crack or in worst-case scenarios explode with great force - capable of damaging other work within the kiln chamber. It is therefore important that a first or biscuit firing is carried out gently. A good general rule of thumb is to fire no faster than 100°C (212°F) per hour. In my workshop I will usually programme biscuit firings in the electric kiln to soak at around 90°C (194°F) for several hours in order to allow any residual moisture to escape. The length of soak time will depend on how thick the work is or how dense the clay. A more open body such as a crank or raku clay, which contains a reasonable amount of grog, will allow the moisture to escape more readily than, for example, porcelain or terracotta. These clays, being less open, will take longer to give up their moisture content. Grog tends to be an umbrella term which describes any relatively course refractory material. For example: sand, molochite, ground fire brick, cordierite, chamotte or a mixture of these materials which when added to a clay body, open the body up and promote more even drying. The introduction of grog will add texture to the clay surface when worked and give the clay more building strength, it will also have the added effect, due to it having been already fired, of reducing the overall shrinkage of the clay to which it is added.

The next stage of the firing is the climb between 90°C and 600°C (194–1112°F). It is through this stage that the majority of the chemical changes involved in the

Holding a visibly dry pot against a window to check for moisture. *Photo: Author*

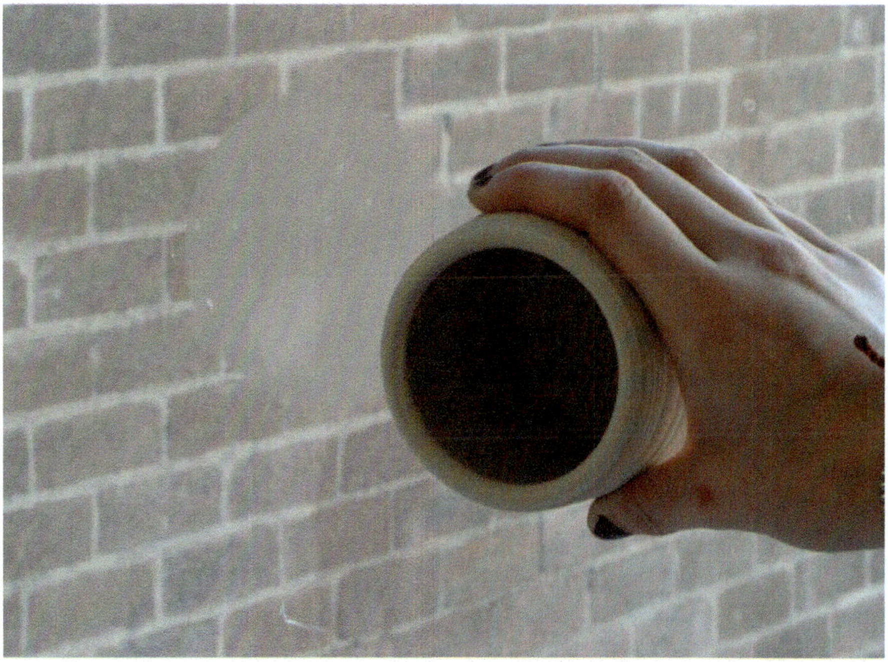

transformation from clay to ceramic occur, generally referred to as the ceramic change. Between 90°C and 300°C (194–572°F), any organic matter still present in the clay is burnt away. If this process is carried out too fast in low-firing clays, bloating may occur due to the surface of the clay becoming vitrified before all the organic material has burnt out. Unable to escape, the gases cause the softened body of the clay to bubble. Also during this stage of the firing, chemically combined water associated with the ceramic oxides at a molecular level is driven off. At around 700°C (1292°F) the clay particles fuse. This is known as sintering and if a biscuit firing were to be stopped at around this temperature then the resulting material would be ceramic, albeit very soft ceramic. In order for it to become strong enough to be handled and glazed, the body must be fired higher.

Silica within the clay also goes through changes in the heating process. This is called the quartz inversion and occurs at approximately 220°C and 573°C (428–1063.4°F). Silica, in the form of quartz or flint, goes through a reversible change which causes an approximately 3% expansion of the work. This expansion happens very fast, as silica crystals are converted to Crystobalite and mullite. If the quartz inversion is ramped through too fast, the expansion (and equally contraction) can cause work to crack, a ceramic fault known as *dunting*, which will be looked at in more depth in another chapter (see p.113).

Generally speaking, there are two commonly used biscuit firing temperatures, referred to as low biscuit and high biscuit. If the subsequent work will be glazed to stoneware temperatures then a biscuit firing of between 960°C and 1000°C (1760–1832°F) (low biscuit), is preferable. At this temperature the ceramic is fairly hard and strong, but still porous enough to accept glaze readily. If the final glazing temperature is to be within the earthenware glaze range, then it is preferable to biscuit fire to around 1120°C to 1160°C (2040–2120°F) (high biscuit). As the glaze firing is likely to be around 1040°C to 1080°C (1904–1976°F) the higher biscuit firing gives the body more strength than if it were fired only to the glaze maturation temperature. In earthenware biscuit firings the temperature can also have an influence on the ultimate glaze fit. It should be noted that a dense high bisque body will be slightly harder to glaze; due to the higher temperature firing, the body will be a little more vitrified and therefore less porous, meaning it will not absorb the glaze so readily. A solution to this problem is to warm the work slightly before glazing.

MONITORING THE FIRING

While a firing is in progress, it is important to be able to observe both the temperature inside a kiln and also what effect that temperature is having on the ceramic oxides present in the clay and glazes. There are several ways of doing this: the most basic electric kiln control may only be an energy regulator dial mounted to the exterior of the kiln. This will usually be calibrated in either percentages or 0–10; either way, the higher the number to which the dial points, the more energy is being run to the elements and therefore the hotter the chamber will get. In the early stages of a firing the dial may be placed at 30% for 1 hour and then moved to 40% for 1 hour, continuing this incremental increase until the kiln is at red heat, at which point the dial may be turned to full until the top temperature is reached and the kiln switched off.

Many more recent electric kilns now come with pre-installed programmers, either analogue or digital, which allow the user to both set the parameters of the firing and get a real-time display of the temperature within the kiln chamber. Controllers

BELOW LEFT: Double energy regulator to control energy input to top and bottom elements of an older style electric kiln. *Photo: Author*

BELOW RIGHT: Analogue temperature indicator is connected to a thermocouple to read kiln temperature. *Photo: Author*

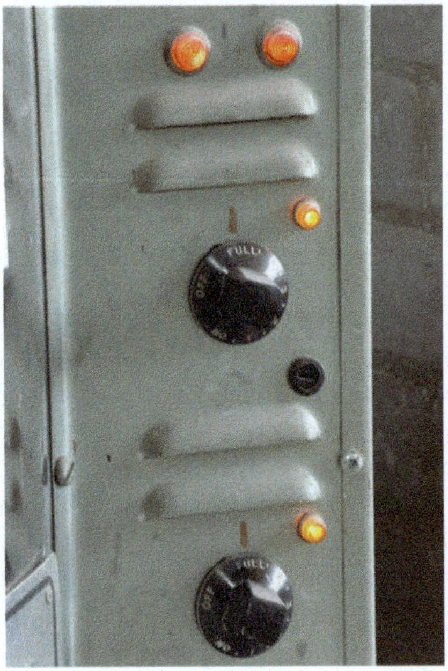

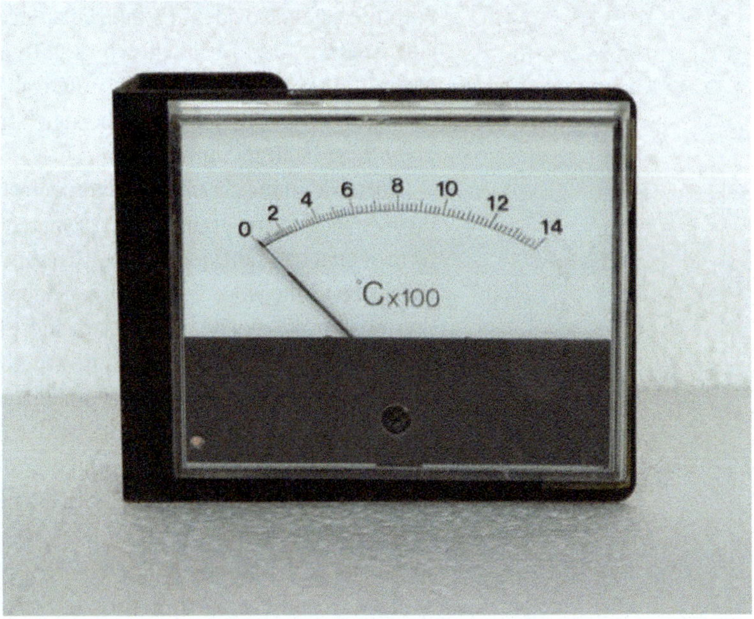

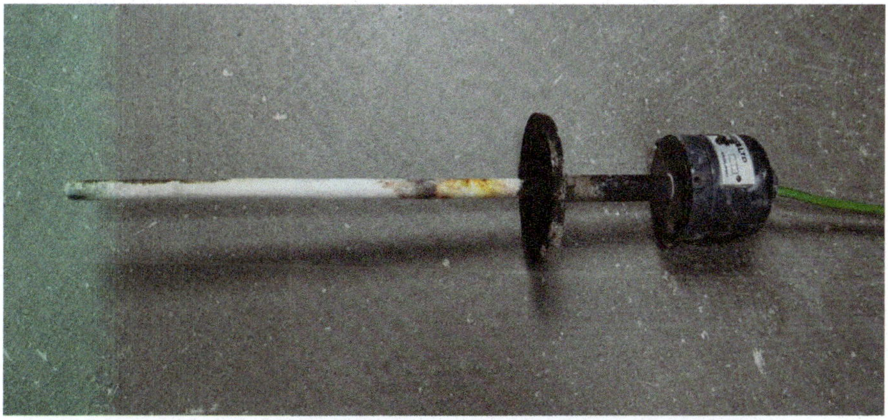

Thermocouple made from two different metal filaments. Type K thermocouples are alumel and chromel. Type S thermocouples are platinum and platinum with 10% rodium. A small milli-voltage is generated from the welded joint between the two metals. This milli-voltage changes degree by degree as the kiln gets hotter/colder. *Photo: Author*

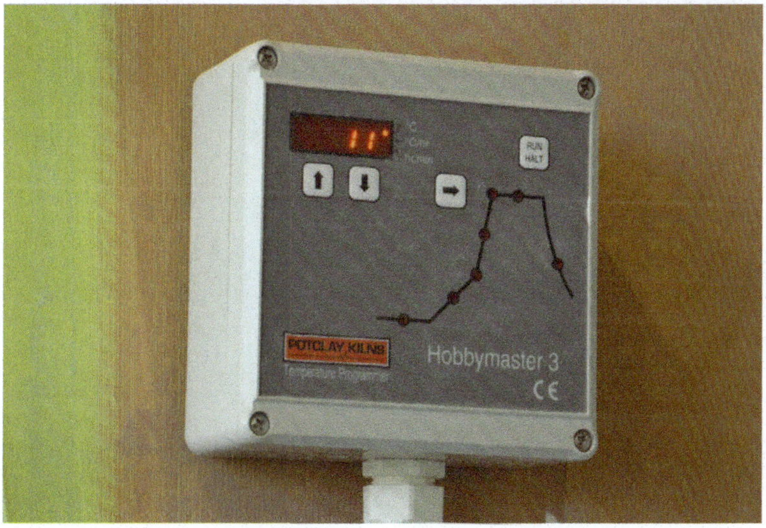

Twin ramp digital kiln controller linked to a thermocouple placed through the kiln wall. Two temperature points can be set. The controller reads the temperature within the kiln through the thermocouple and inputs energy into the elements to achieve the programmed temperature climb. *Photo: Author*

range in their functionality; at their most simple they allow the user to set the top firing temperature and have a limit function that will turn the kiln off automatically when it reaches the desired heat. At the more complex end of the spectrum, a programmer can possess multiple user programmable ramps, which can be linked together to create very complex firing schedules. All kiln controllers work by being connected to a thermocouple, usually a porcelain sheathed bi-metallic probe that extends into the kiln chamber and is connected to the kiln controller, either directly or by a compensating cable.

Firing terms frequently seen on kiln controllers

Programme

This usually describes a complete firing cycle. A cycle can be pre-programmed, e.g. high-biscuit firing, low-biscuit firing, stoneware glaze firing, etc. Or, user programmable, which allows the user to input the desired ramp times and temperature set points.

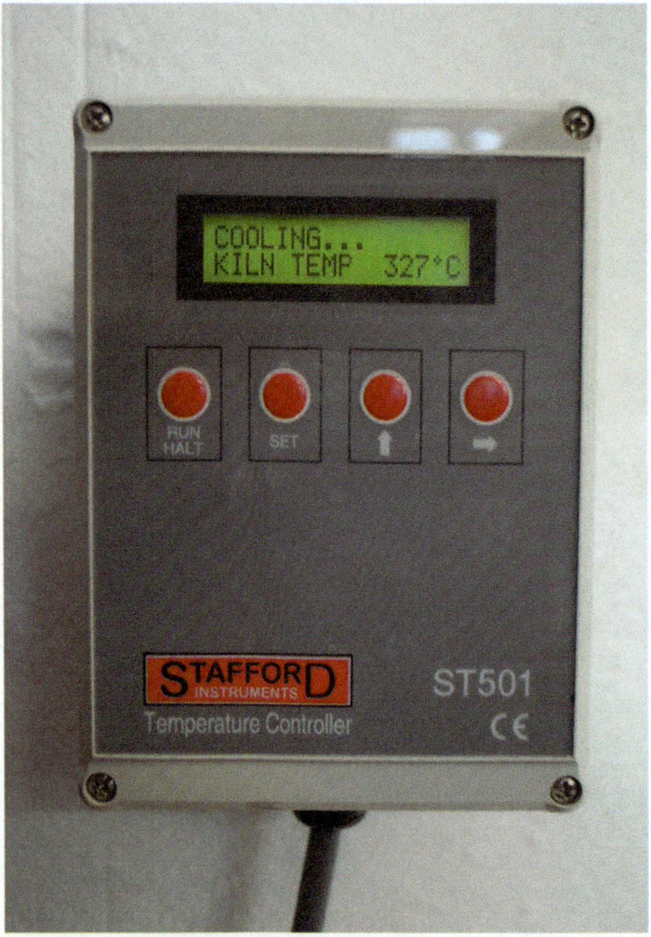

Multiple ramp digital controller. This type of controller can contain 20 user programmable programmes containing 10 segments each, allowing the potential to link together 200 temperature set points. Needing this degree of control is very rare within studio ceramic practice but the ability to control the cooling of the kiln with the controller can be very beneficial when carrying out certain specialist firings such as macro crystalline glaze firings. *Photo: Author*

Segment

This describes the separate programmable sections that can make up a complete programme. For example, segment 1: 0–600°C (0–1112°F), at 100°C/hour (212°F/hour). Segment 2: 600–840°C (1112–1544°F) at 125°C/hour (257°F/hour).

Ramp

This describes the number of degrees per hour that a kiln will climb to a specified temperature.

Set point

This refers to fixed temperatures that can be programmed into the kiln controller, e.g. in the above example, 600°C and 840°C (1112–1544°F) would be the set points.

Soak/dwell

The ability to hold a specified temperature for a period of time; useful for drying work at the low end of a firing, or maturing clays and glazes at the top end of a firing.

HEATWORK

There are several other ways of monitoring the progress of a kiln's firing. While kiln controllers can be useful as an indicator of temperature within a kiln and also timed to reach set points, what they cannot measure, and what is arguably a more important aspect of kiln monitoring, is the action of that temperature and time on the ceramic materials in the kiln pack. This is known as 'heatwork' and there are several ways of measuring it.

Bullers Rings

Bullers Rings are formulated using similar materials to those used in glaze composition. They contract over the duration of a firing. By measuring the ring in a dedicated micrometer after the firing or removing and allowing to cool through the firing, an accurate understanding of the degree of heatwork to which the work is subjected can be ascertained. This type of heatwork recording tends to be used rather more in an industrial context. Arguably the studio ceramics equivalent would be pyrometric cones, and in the UK these are more commonly used.

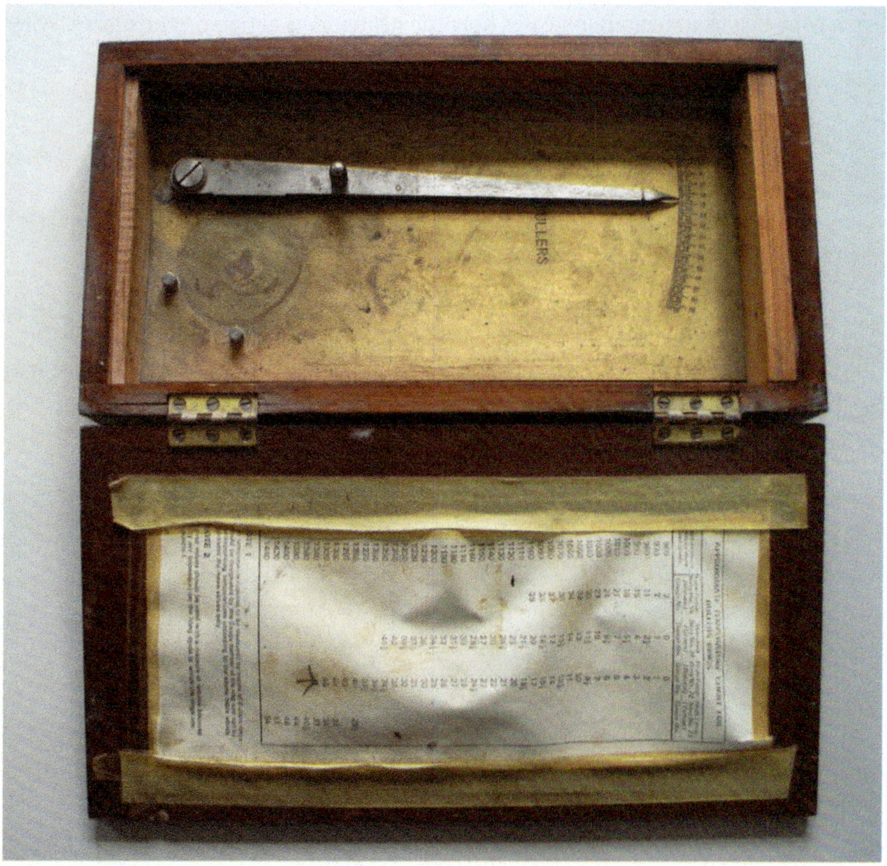

Bullers ring measurer. A ring is removed from the kiln through a hole in the kiln wall, cooled, and then placed on the scale and the pointer lowered until it is touching the ring. The temperature can then be read from the end of the pointer. *Photo: Author*

Pyrometric cones

Pyrometric cones are produced by several manufacturers although the most common are produced by Orton, and are formulated in a similar way to Bullers Rings, but have the added advantage of giving a real-time indication of heatwork within the kiln chamber as they can be placed in a holder and viewed through the kiln's spy hole during the firing. Cones are numbered to indicate the temperature at which they are designed to melt, starting as low as 022 (600°C/1112°F), and going up into the 30s. Realistically though, within studio ceramic practice cones within the range of 07 (980°C/1796°F) to 10 (1300°C/2372°F) are most commonly used. Most Pyrometric cones come in three types - large, small and self-supporting. They can be placed either in a dedicated holder, simply supported by a small coil of clay wrapped around the base, or multiple cones placed in a bed of clay (crank clay tends to be best for this as it is more open and readily allows any moisture to escape at the beginning of the firing). Using multiple cones means that you can have a higher and lower temperature cone on either side of your ideal temperature, to give more information.

It is important to ensure that the base of a cone is flat on the kiln shelf. If the cone is too upright, it will fail to fall at the right time. Once dry, the cone pack can be placed in front of one of the kiln spy holes and watched through the firing. As each cone begins to bend it indicates how the heatwork is acting on the glazes. When

BELOW, TOP ROW LEFT: A Pyrometric cone pack is positioned in front of a spy hole in the kiln wall so that it is visible throughout the firing. Or, it can be placed blind to confirm that a top temperature has been reached post firing.

RIGHT: Orton cone 3 bending in a stoneware reduction firing.

BOTTOM ROW, LEFT: Cone pack after firing: cones low.

RIGHT: Cone pack after firing: cones high.

Photos: Author

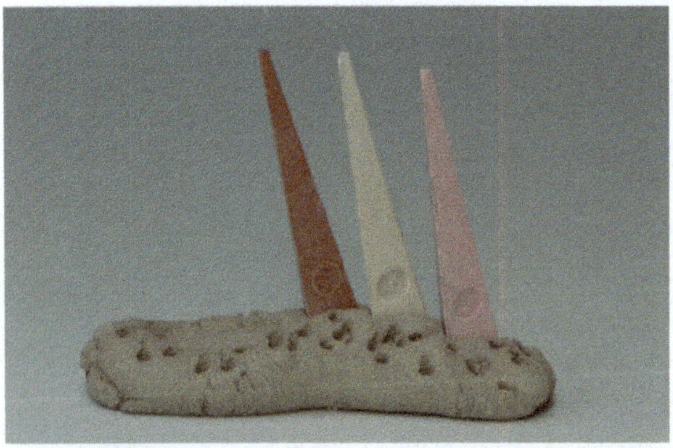

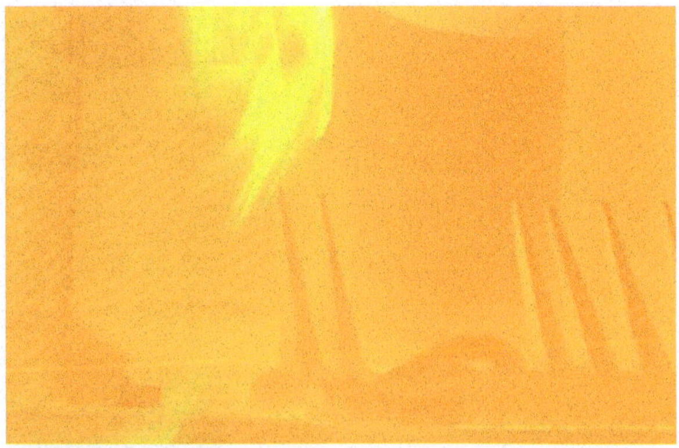

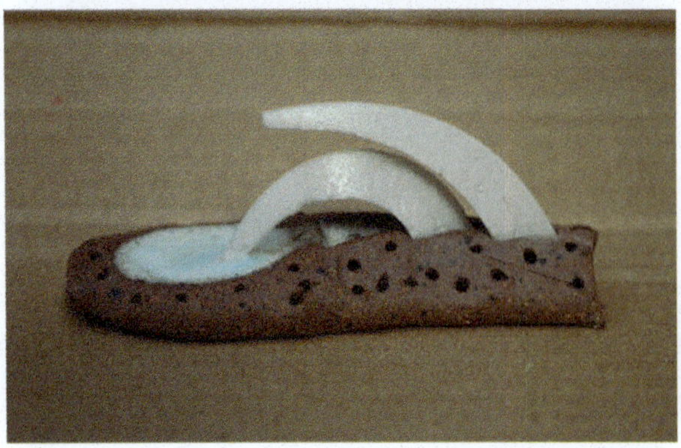

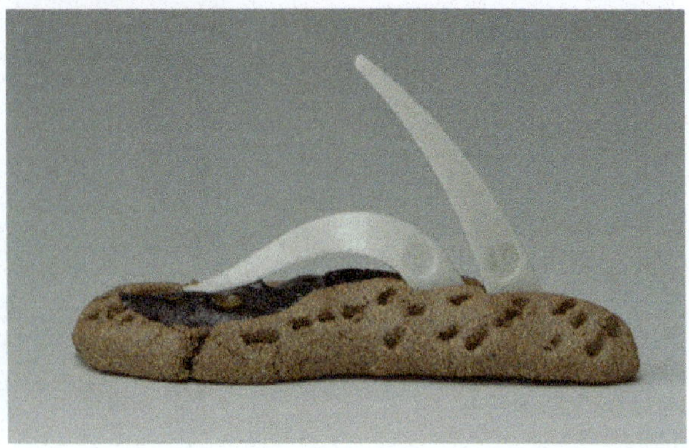

viewing cones there is still a degree of interpretation that is a subjective choice. Some individuals consider a cone to be over when the tip is bending; others choose to allow the cone to bend until the tip touches the shelf before considering it over. It is all down to personal preference and experience. When the firing is over and the kiln has cooled, the cones can be kept as a record to inform subsequent firings.

Pyrometric bars

Pyrometric bars are formulated in a similar way to pyrometric rings and cones, but come as small, square, rectangular or triangular section bars. These can be placed in holders that are supported either end. As the kiln reaches the appropriate point, the bar will bend.

Bars can also be used in kiln sitters. A kiln sitter is a device designed to switch off an electric kiln automatically. The bar is held in a sprung jaw system; when the bar melts, the jaws close and release a switch on the exterior casing of the kiln, which in turn disconnects the power going to the elements.

FIRING SCHEDULES

To a certain degree, the firing schedule one follows is down to personal preference and experience, as well as the clays and glazes being used. It is a good idea to record firings so that there is always a guide as a reminder. As stated earlier, however, there are some precautions that should be applied to all firings – mainly that the ramps between around 200°C and 600°C (392–1112°F) are passed through cautiously both on the way up and on the way down due to quartz inversions. This especially applies to denser clay bodies.

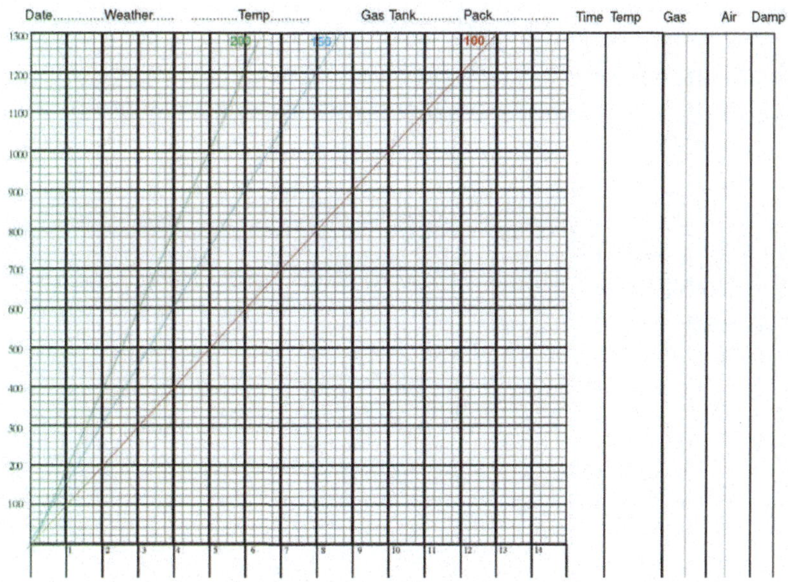

Firing graph for Alistair Young's kiln firings. *Photo: Alistair Young*

Firing schedules for multi-ramp programmers

Standard low-biscuit firing

Segment 1:
Set point: 600°C/1112°F
Ramp: 100°C/hour (212°F/hour)

Segment 2:
Set point: 980°C/1796°F
Ramp: 150°C/hour (302°F/hour)
Finish

Low-biscuit firing for thicker work

Segment 1:
Set point: 90°C/194°F
Ramp: 50°C/hour (122°F/hour)
Soak: 4 hours

Segment 2:
Set point: 600°C/1112°F
Ramp: 100°C/hour (212°F/hour)
Soak: 30 mins

Segment 3:
Set point: 980°C/1796°F
Ramp: 150°C/hour (302°F/hour)
Soak: 30 mins
Finish

Stoneware glaze firing

Segment 1:
Set point 600°C/1112°F
Ramp 100°/hour (212°F/hour)

Segment 2:
Set point 1260°C/2300°F
Ramp Full power
Soak 30 mins
Finish

Macro crystalline stoneware glazes (slightly more complex firing)

Segment 1:
Set point 1260°C/2300°F
Ramp 100°C/hour (212°F/hour)

Segment 2:
Set point 1120°C/2048°F
Ramp: 50°C/hour (122°F/hour)

Segment 3:
Set point 1260°C/2300°F
Ramp full power
Finish

The author: stoking the kiln.
Photo: Author

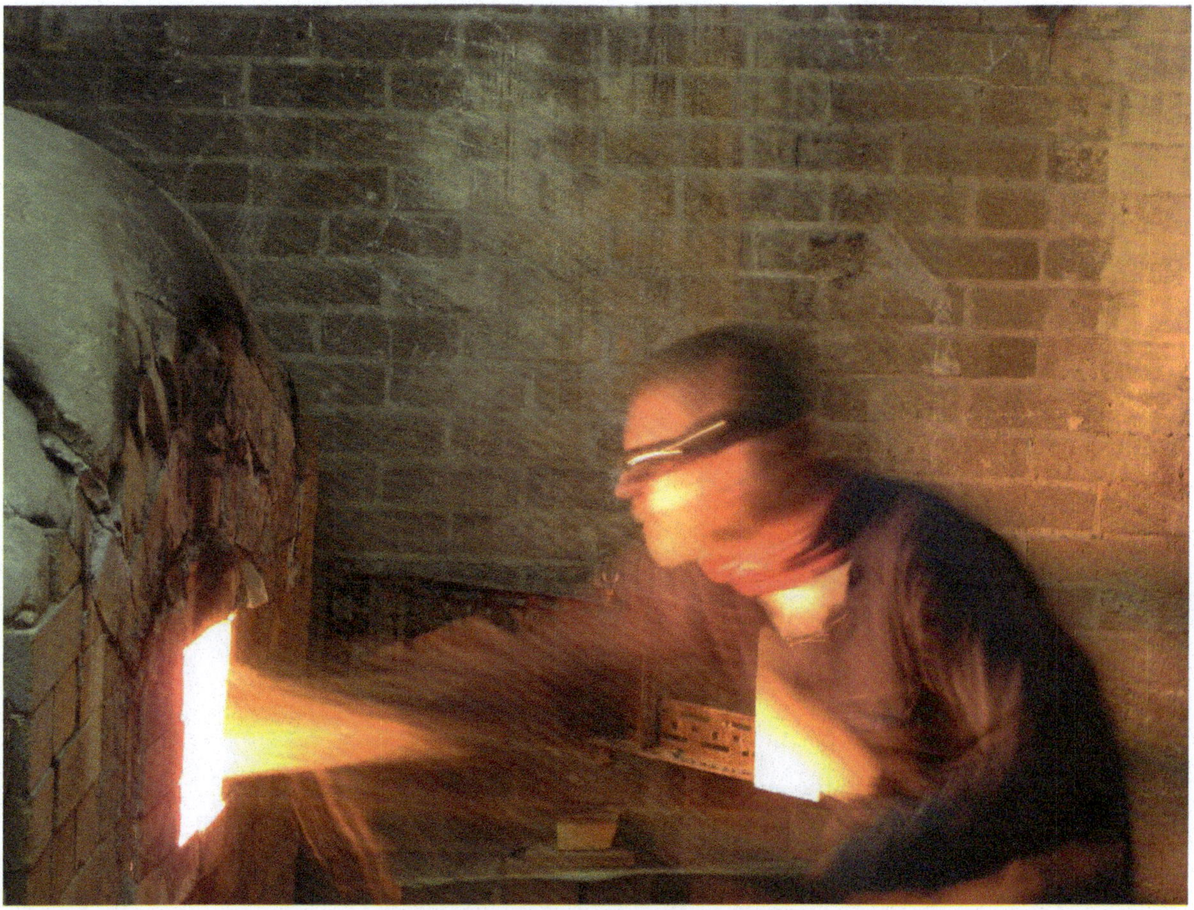

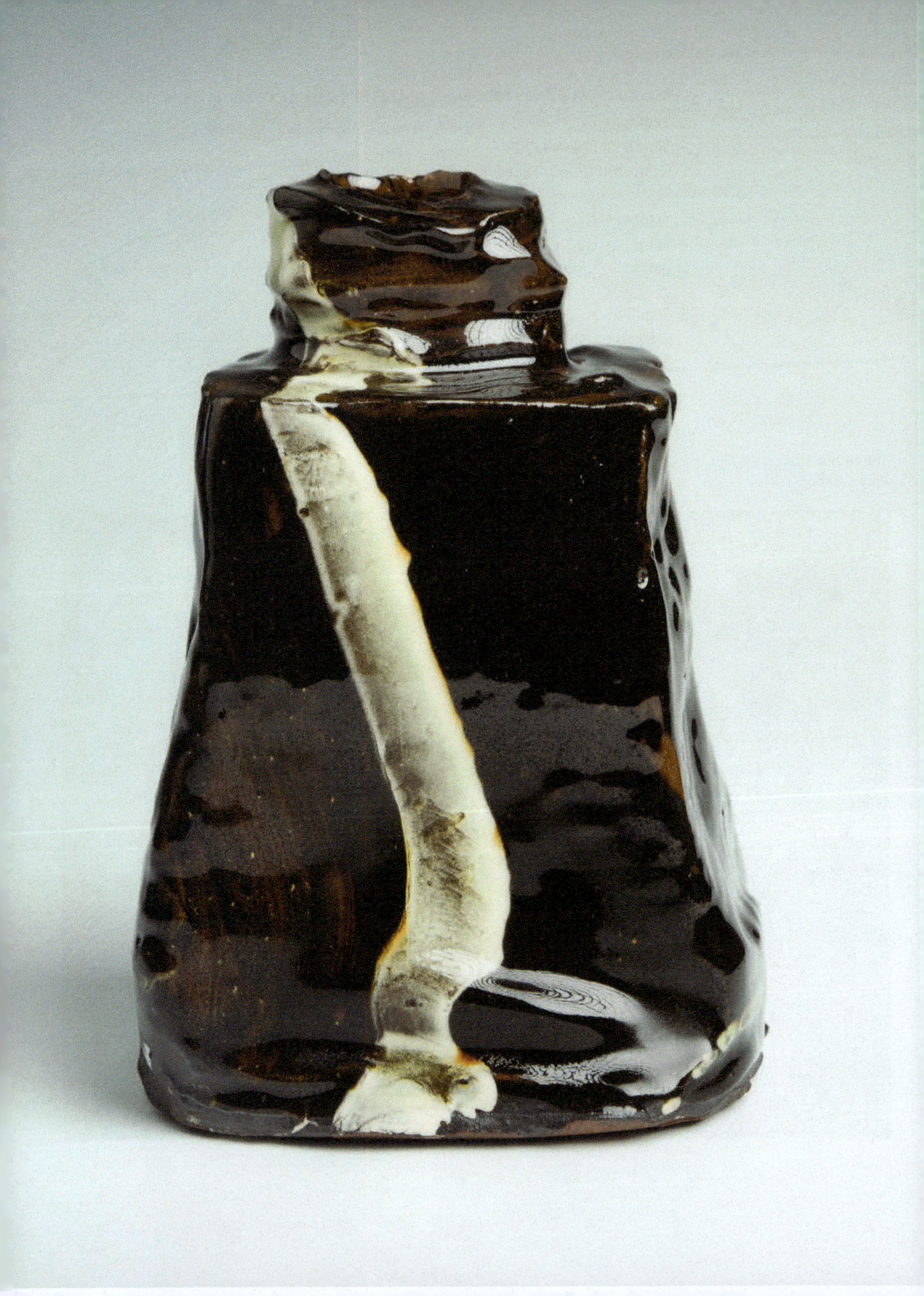

2

Types of kilns and packing

The earliest firing structures were simply shallow hollows in the ground in which clay objects were placed and covered with brush and wood. When lit, with careful management of the burn, temperatures within the pit reached high enough to produce the ceramic change. As kiln understanding increased, in order to achieve more efficiency and generate higher temperatures, an adobe roof was added. Arguably this would have comprised a mixture of clay, straw and possibly animal dung, not only giving a structurally strong cover but as the straw combusted in the firing, also creating voids in the clay which allowed the roof to act as an insulator.

Kiln technology has moved on from this considerably, and there now exist many types of kiln which can be used to generate many different ceramic surfaces. In the late 20th and early 21st centuries many of these old kiln technologies are being revisited by ceramic artists and potters wishing to explore their indirect influence on ceramic surfaces. This will be discussed more widely in Chapter 5.

ELECTRIC KILNS

An electric kiln, as its name implies, is a kiln which uses electricity to heat elements housed within the walls of the kiln. The heat given off by these elements raises the temperature within the chamber until it reaches the desired point. Electric kilns are constructed using high-temperature insulation bricks (HTI), housed within a steel frame and casing. The HTI hot face brick is backed by ceramic fibre board for added insulation. In older kilns over 30 years old, it is worth noting that asbestos was often used for this secondary heat-proofing layer and great care should be taken when carrying out any maintenance work.

The amount of power going into the elements is controlled by either a kiln controller, or is completely manual, achieved by turning up the energy regulator at regular intervals. Most kilns without controllers will have an energy regulator dial which will be graded 0–100%, increments of 10 (from 10–100), or in older kilns, low, medium and high.

Differing makes of electric kilns may have slight variations in element configurations. Some may simply have elements housed in channels running down the side walls, others may also have elements housed in the floor and door. How much heat the elements are capable of emitting will depend on the number of windings and the gauge of the wire used. A smaller gauge wire with more windings will deliver the same heating power as a heavy gauge wire with fewer winds, etc. Elements are mainly made using Kanthal A1 wire and are capable of working up to about 1400°C (2552°F),

Cut Bottle, Dylan Bowen. Electric kiln-fired earthenware, 17 x 10 cm (6½ x 4 in). *Photo: Ben Ramos*

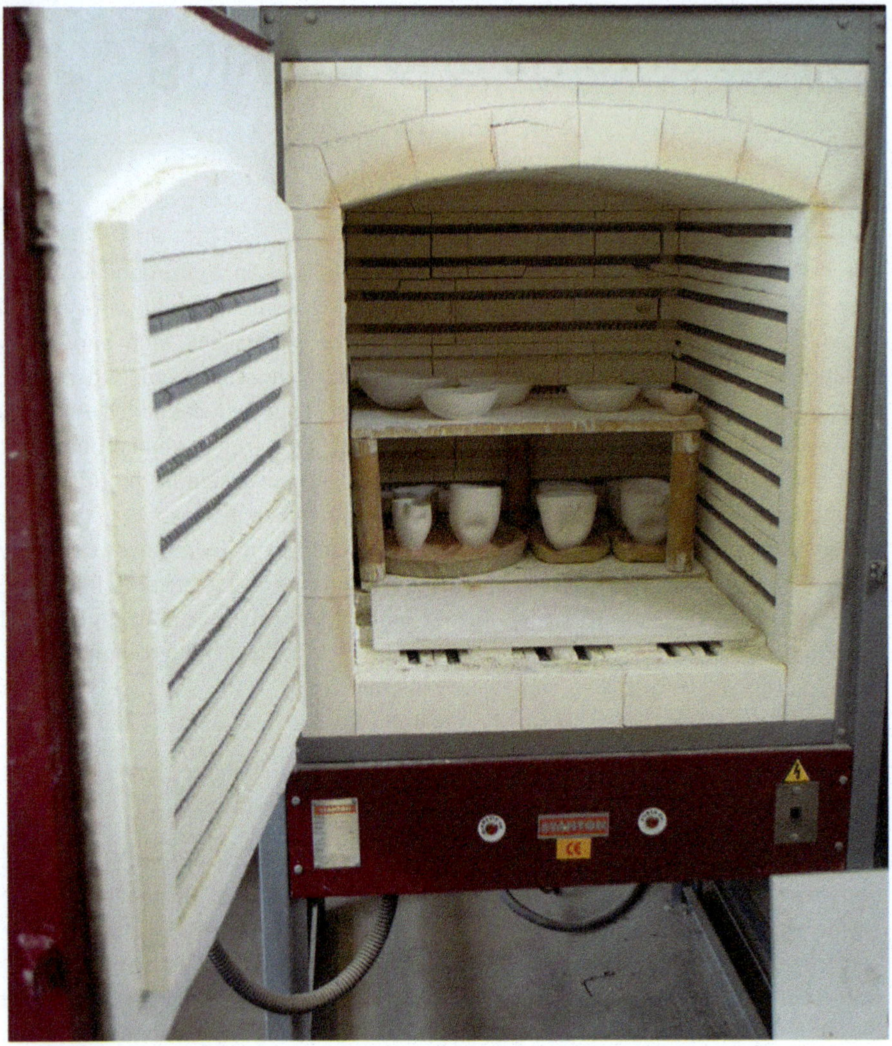

Electric kiln with elements housed in walls, door and floor. *Photo: Author*

although if wanting to work up to this temperature one would need to ensure that the grade of brick used was capable of working at those temperatures.

All the elements within the kiln connect up to make a circuit and the period of time that energy is allowed into the elements is governed by a contactor. This is the sound which can be heard while an electric kiln is firing; a clunking sound as the contactor switches in and out. Electric kilns heat by allowing energy to the elements in increasingly long periods. A hypothetical example could therefore be: with the kiln energy regulator set at 30%, in any given minute, energy will be heating the elements for 20 seconds. If the energy regulator is then increased to 50%, energy will be going into the elements for 30 seconds; as energy is heating the elements for longer the internal kiln temperature will rise accordingly. When the energy regulator is set to 100%, power is constantly going to the elements and the kiln temperature will continue to climb until the desired top temperature is reached.

Oxidised slipcast parian ware: from the series *Trans Forms Plus (Bastards' stories)*, Monika Patuszynska. Fired in an electric kiln to 1260°C (2300°F), height: 17 cm (6½ in). *Photo: Monika Patuszynska*

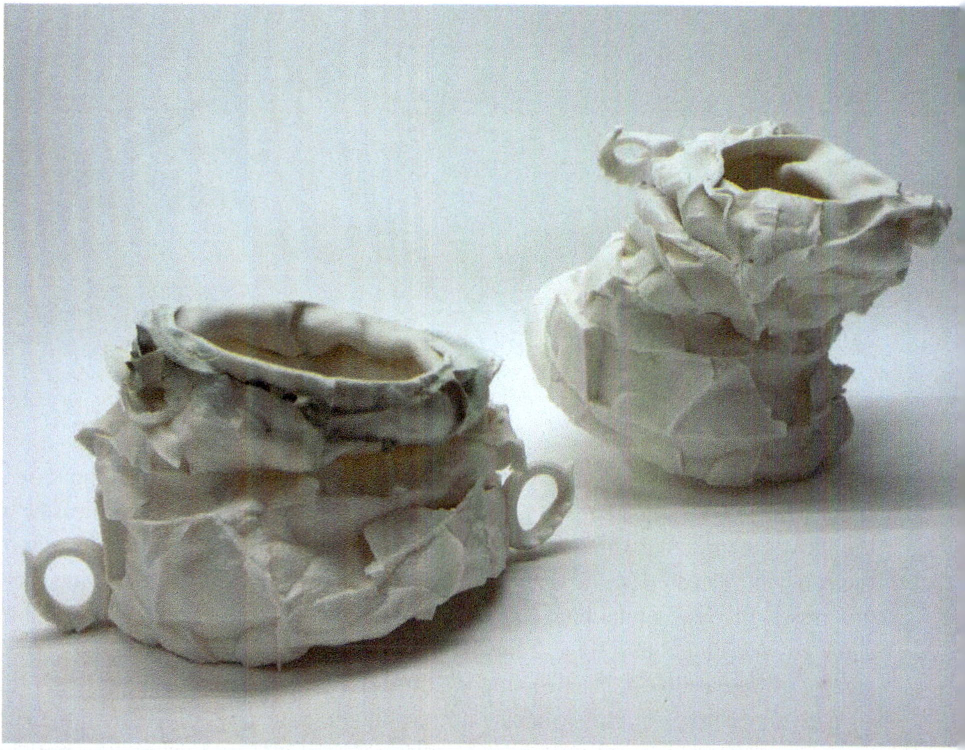

BELOW: *Wash line (*group of demitasse), Thomas Aitken and Kate Hyde. Oxidised stoneware fired in electric kiln to cone 6 (1200°C/2192°F), 9 x 6 cm (3½ x 2½ in). *Photo: Thomas Aitken*

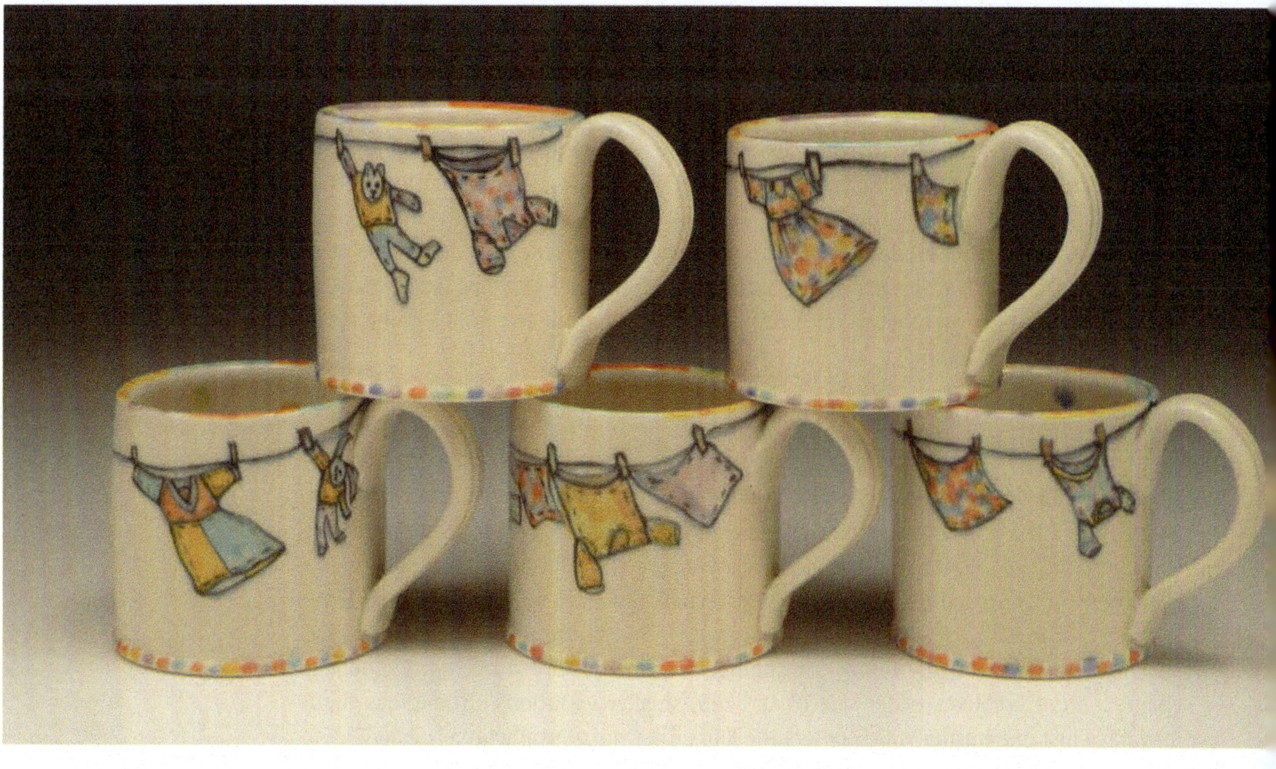

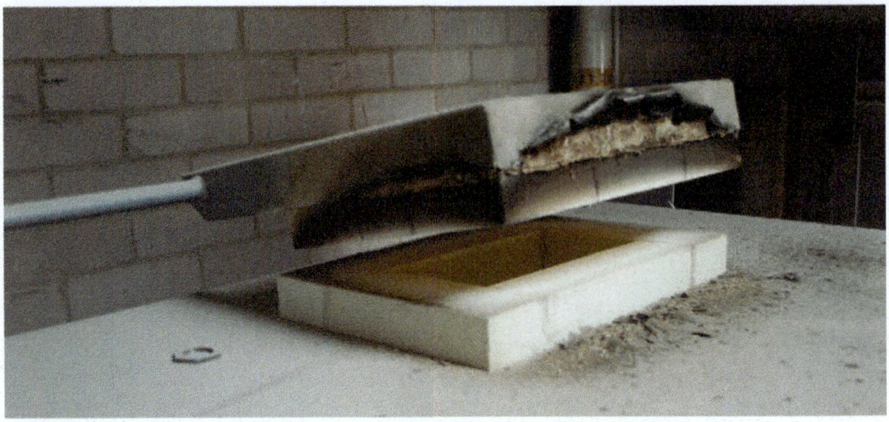

Adjustable top vent on a 16 cu. ft electric kiln: the bar running through the back of the vent lid is attached to a handle for ease of opening and closing. The vent is left open for the first 300–400°C (572–752°F) climb of the firing and then closed. At the end of the firing it may be opened a little to facilitate slightly faster cooling of the chamber. *Photo: Author*

Many electric kilns will have a vent in the roof, this may be a simple hole or it could be connected to a handle so that it can be opened and closed more easily. In the most sophisticated kilns, vents may be connected to a motor and ultimately to the kiln controller so that the vent will be controlled automatically as the firing cycle progresses. The vent would tend to be left open during the early stages of a firing to allow any moisture and residue from the burn out of organic material in the clay to escape; after the kiln has reached around 600°C (1112°F), the vent can be closed.

With an electric kiln, it is generally not possible to have any control over the kiln atmosphere. Please see reduction in electric kilns (Chapter 4) for further information.

FUEL-BURNING KILNS

Fuel-burning kilns fall into three main categories of design: downdraft, updraft and crossdraft. Each of these designs offers certain variables to the work which is fired within them. Probably, the most consistent with regards to evenness of heat distribution would be the downdraft kiln, followed by the updraft and finally the crossdraft kilns. No kiln design is any better than the other but one may be more appropriate than another dependant on the type of work that is required to be fired, or the surface marks desired.

Downdraft kiln

Details in kiln design may vary slightly depending on what fuel is to be used: the majority of kilns are either sprung arch or catenary arch in construction, however the overruling principles are the same. Catenary arch kilns are a self-supporting continuous arch structure, whereas sprung arches have the roof 'springing' off the top of vertical walls; this type of design requires a metal frame support to stop the arch from spreading at the top of the chamber walls. The heat source enters the kiln chamber from fireboxes either side of the kiln (smaller variations, usually gas fired, may have just one burner which enters the kiln either horizontally or vertically).

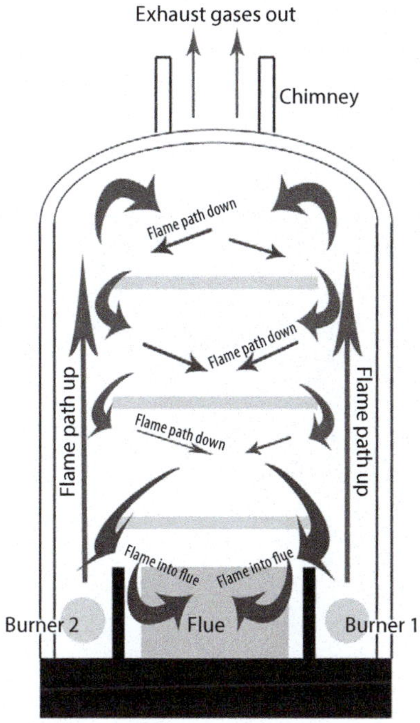

ABOVE: Diagram of the flame path formed within a downdraft kiln: heat rises to the top of the chamber and is then drawn down through the pack to the flue and out through the chimney.

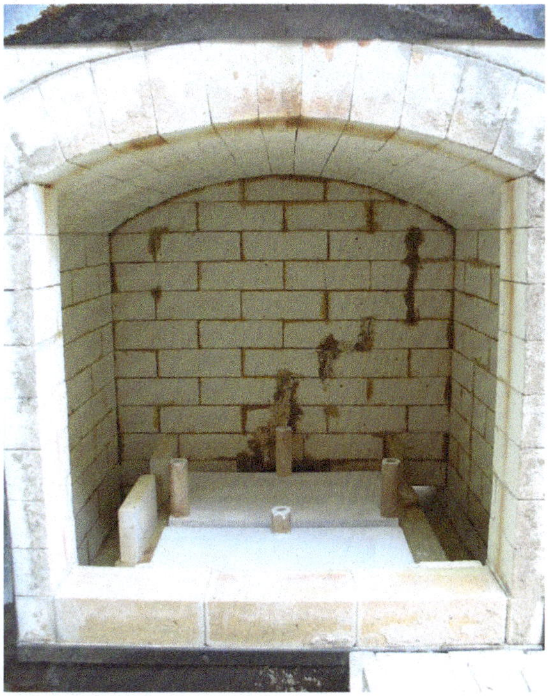

ABOVE: Interior of a Antje Ernestus's downdraft gas-fired kiln. *Photo: Antje Ernestus*

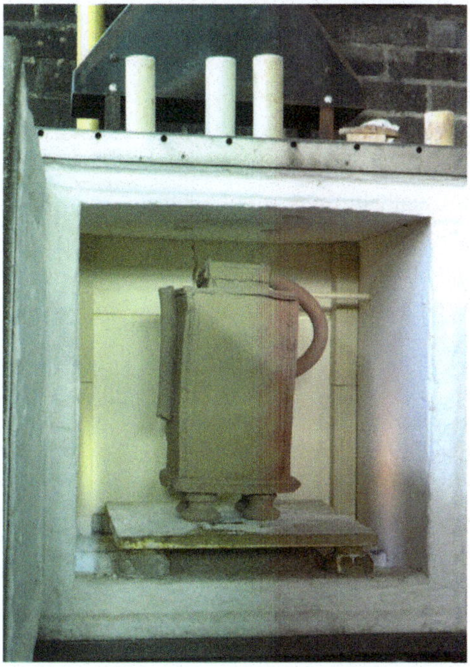

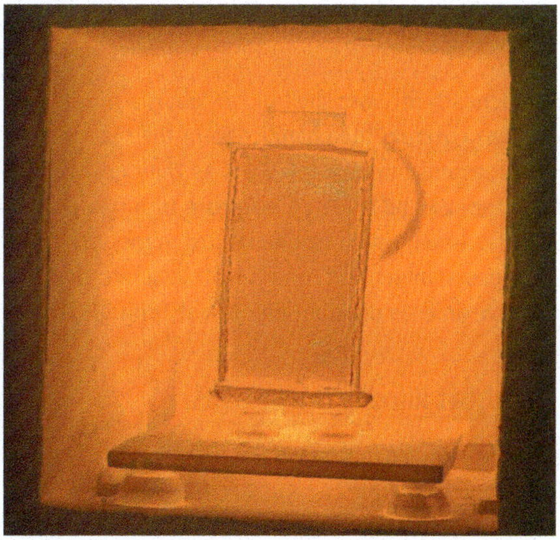

LEFT: Small downdraft gas-fired kiln with piece in situ.

ABOVE: Inside a small downdraft kiln at 1260°C (2300°F).

Photos: Author

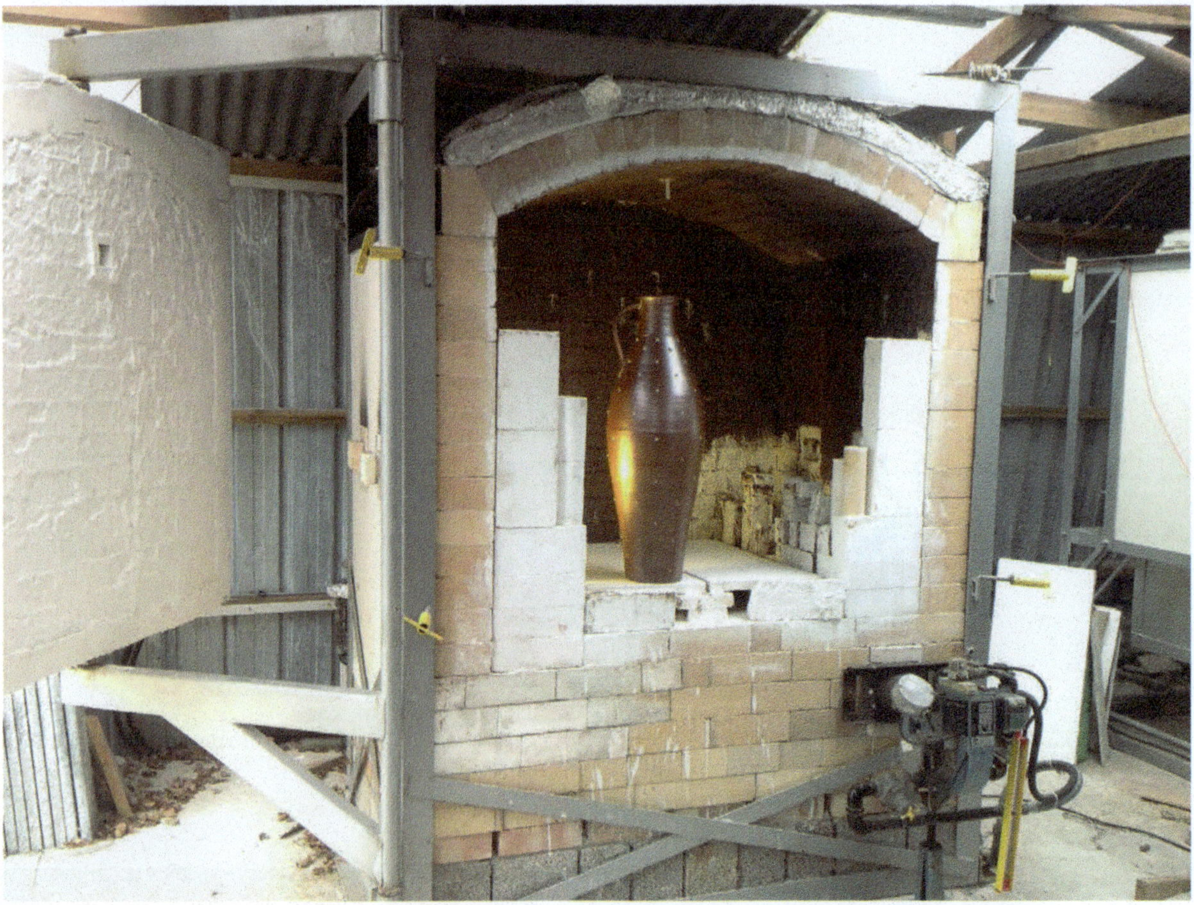

The interior of Alistair Young's medium-sized gas-fired downdraft kiln. *Photo: Alistair Young*

Beneath the base shelf of the kiln is a flue which enters a chimney in the back wall. Flames entering the chamber from the burners rise and are then drawn down to the flue passing through the work as they go heating the work up. The draw from the chimney can be adjusted by means of a damper which is usually in the form of a kiln shelf or brick which can be moved across the chimney and restricts the flow of heat and gases, thereby altering the atmosphere within the chamber. This type of kiln tends to give an even distribution of heat throughout the kiln chamber.

Updraft kiln

This type of kiln design utilises rising heat to heat the chamber. Fireboxes are generally positioned around the outside wall of the kiln with the fire mouths entering the chamber beneath the stacking space. On larger kilns the entire kiln floor may be perforated to allow in the heat from the fireboxes. Heat then rises through the work and out of a chimney on the top of the chamber.

The most common examples of this type of kiln are the bottle ovens that were so prevalent in Stoke-on-Trent in the 17th and 18th centuries. Updraft kilns tend to

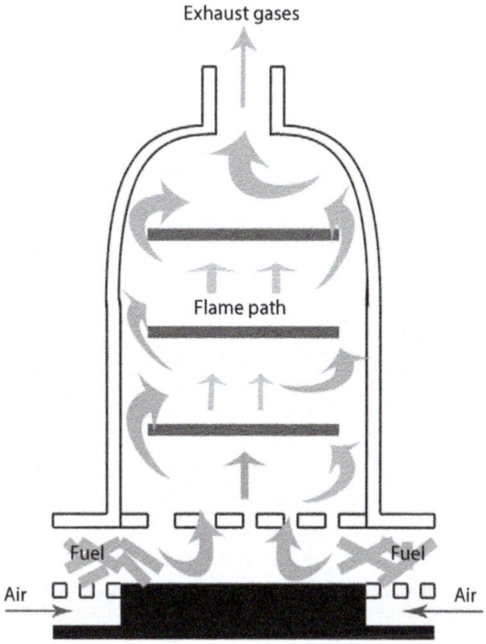

heat slightly unevenly and this disadvantage was utilised in the Potteries by placing specific work in specific areas of the kiln dependant on the temperature required to mature the clay and glaze. Most work was fired in saggars to protect it from sulphur which was emitted from the coal when burnt. Saggars were stacked one upon another with high-fired work such as bone china being stacked in the parts of the chamber which got hotter and lower-firing earthenware being stacked in cooler areas. In this way, the whole of the kiln chamber could be utilised with no wasted space.

LEFT: Diagram of the flame path within an updraft kiln: heat enters the kiln beneath the pack and passes through the work, exiting through a chimney at the top of the kiln.

BELOW LEFT: Disused updraft bottle oven, Stoke-on-Trent. *Photo: Author*

BELOW: Woodfired updraft bottle kiln at Wobage craft workshops. *Photo: Patia Davis*

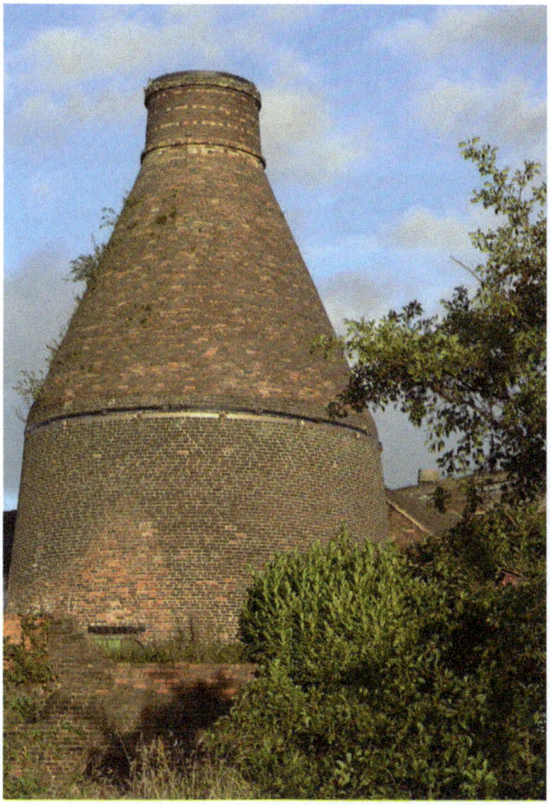

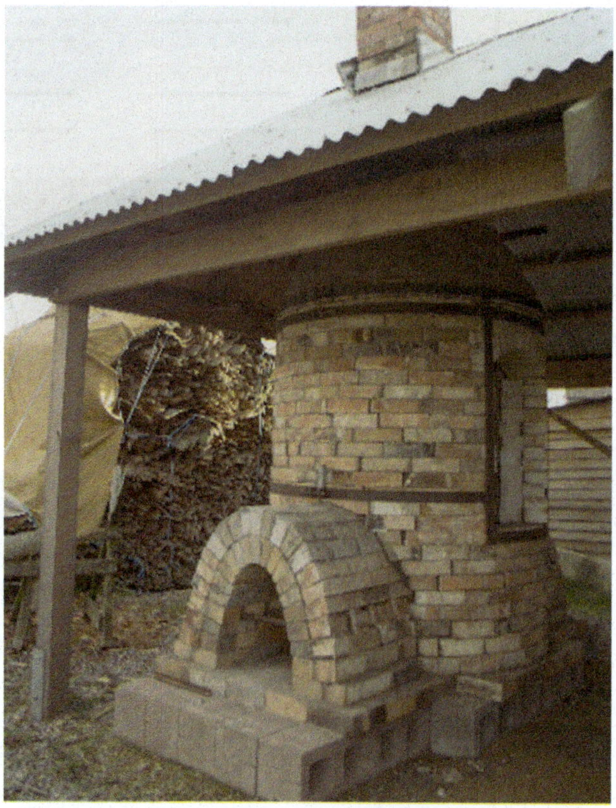

Crossdraft kiln

A crossdraft kiln as its name suggests is a kiln which is heated from one end, with a chimney for exhaust gases at the other. Work is packed into the kiln chamber and heat is drawn through it thus raising the kiln's temperature. As with updraft kilns, this type of structure inherently has an unevenness in temperature distribution, as the work towards the front of the kiln will always tend to be hotter than the work positioned towards the back. However, makers who tend to fire in this type of kiln allow for this, and work is positioned appropriately for the surface finishes required.

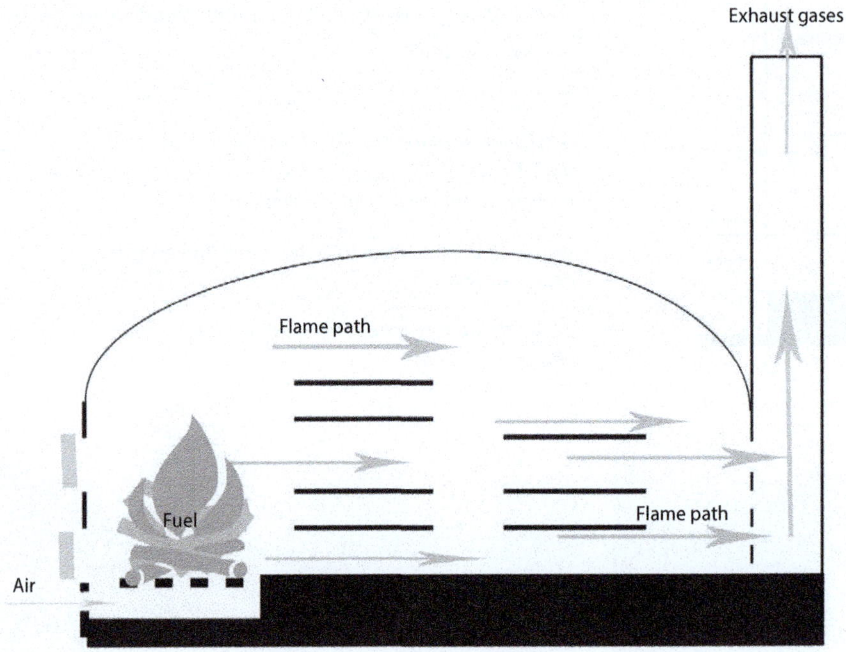

Diagram of the flame path within a crossdraft kiln: heat travels from the flame source at one end of the kiln, through the work, exiting through a chimney.

KILN-FIRING FUELS

Gas

Gas is probably one of the most common fuels used for firing kilns. Both natural gas and propane are readily available. Gas also gives the advantage of being relatively clean and fairly easy to install, either by direct link to a mains supply, or by the use of portable propane tanks. Propane has a higher calorific content than natural gas, and burners will be fitted with different jets depending on the fuel to be used.

Gas-fired kilns can be used to carry out pretty much any firing from a slow biscuit firing to a high-temperature stoneware reduction. One drawback to firing with propane, however, is if a single bottle is used, when the gas level drops the bottle will tend to frost up, which decreases the gas pressure. This becomes more of an issue when carrying out higher temperature stoneware firings, when the extra pressure is

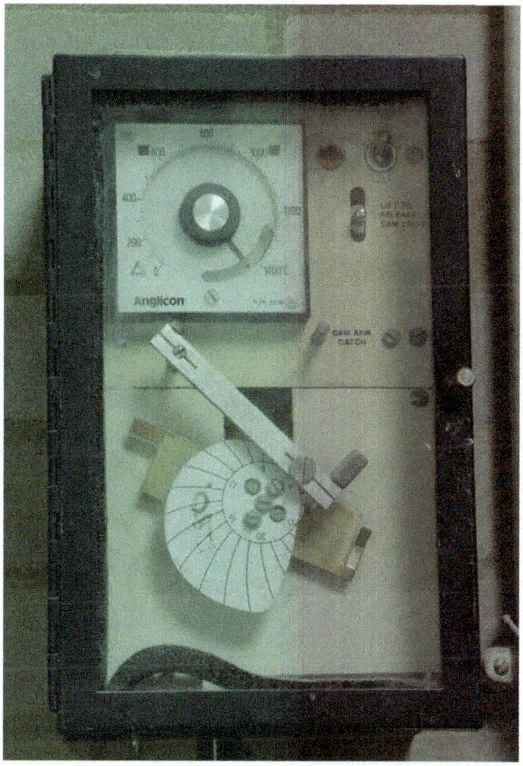

Gas kiln, cam controller. Cam plates (the metal seashell-shaped plate), can be changed to alter the firing ramps, etc. As they rotate, the arm which rests on the cam controls the gas and air input to the kiln. *Photo: Author*

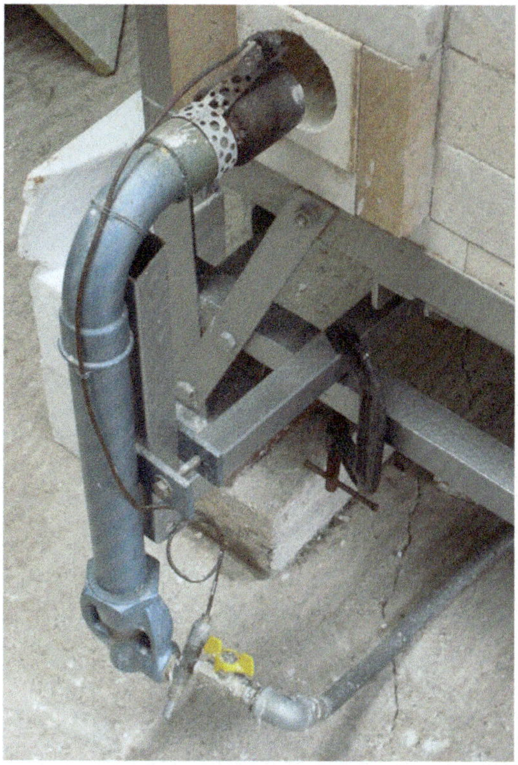

Horizontal gas burner. *Photo: Alistair Young*

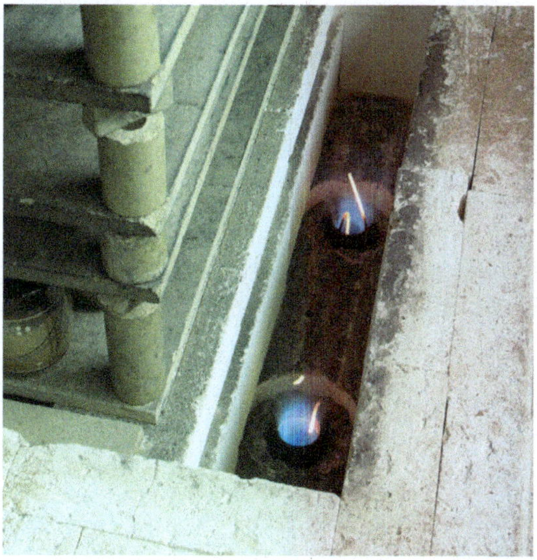

Vertical gas-kiln burners. *Photo: Françoise Dufayard*

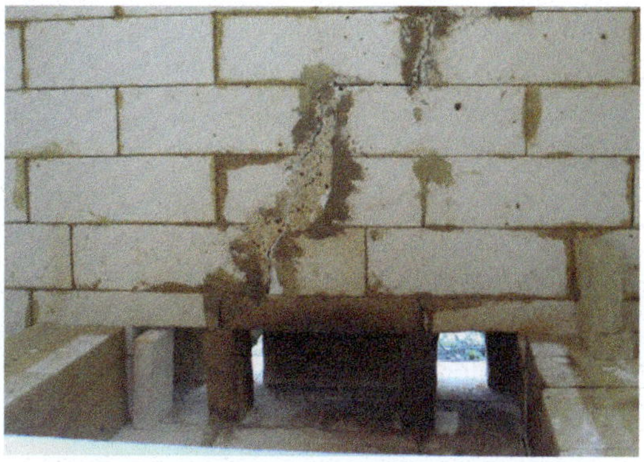

Flue of a gas-fired downdraft kiln. The holes visible through the back wall of the chimney are passive dampers: when closed, the chimney has full draw, but when out, the chimney will pull in some air through them, thereby lessening the pull through the kiln chamber. This will increase the reduction within the chamber. *Photo: Antje Ernestus*

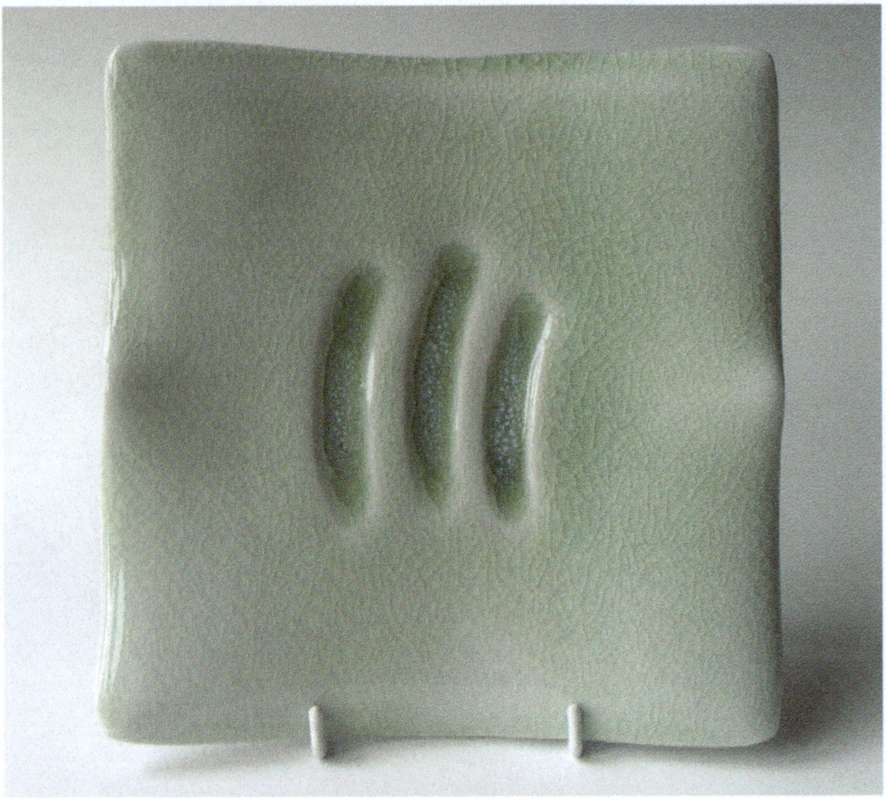

Porcelain platter with blue ash glaze, Antje Ernestus. Fired to Orton 1300°C (2372°F) in her propane gas-fired downdraft kiln, to cone 11, dimensions: 30 x 30 cm (12 x 12 in). The kiln is a straightforward downdraft design with an integral chimney built into the back wall, fired on a pair of propane gas bottles connected to two gas burners (Aeromatic Barter Ltd) set at diagonally opposite corners. *Photo: Antje Ernestus*

useful to gain the higher temperatures. If regular stoneware firings are to be carried out it is worth considering linking two or even three bottles together. When linked, the emptier bottle can be used for the early stages of a firing and switched over to the fuller bottle for finishing.

Gas kilns can be fitted with thermocouples and temperature indicators, often introduced through a hole in the external metal casing and protruding through the HTI brick.

Gas kilns can also be fired using a rotating cam system to regulate the gas flow, however it is generally less common for the whole of the firing process to be automated. Pyrometric cones can be placed in front of spy holes usually top and bottom, and it is by observing the cone bend that dictates when to finish the firing. Gas-fired kilns can be any of the above designs although within the sphere of studio ceramics probably the most common design is that of the downdraft kiln.

Wood

Arguably wood is one of the oldest methods of firing kilns and, historically, has been identified as a major influence on deforestation in some parts of the world, especially during medieval times when localised pottery production was escalating. Today, however, wood is used much less frequently, as gas has become much more popular for the firing of fuel-firing kilns for the reasons stated above. However, there began a resurgence in using wood as a fuel for studio ceramic workshops from the 1970s onwards.

Wood is a variable fuel, different woods can have widely differing calorific outputs so, when firing with wood it is important to understand the relative amount of energy that a particular wood has. It is also a fuel that is being used more widely as it is a renewable resource and can be sourced from sustainable supplies. Unusually, wood is a fuel which can have a direct modifying effect on both the clay bodies and glazes, and this will be discussed later.

The design of a wood kiln, arguably more than any other, requires very particular firebox dimensions. In order for a wood-fired kiln to work, it relies on a balance between the sizes of packing chamber, firebox and chimney. The chimney sucks air in through the firebox or grate, combusting the fuel and producing heat. The firebox should be large enough to ensure that enough air can enter the combustion area, allowing for a build-up of coals and ashes.

It is a common error when building wood-fired kilns, not to allow room for enough air to enter the grate, the consequence of which is that the firebox becomes choked with embers. The kiln relies on the embers built up from combustion to preheat the air being drawn in, which ensures that the kiln is not losing too much temperature when the fuel is placed into the firebox. If the fireboxes require lots of raking out to allow enough air in, then the dimensions are probably slightly out. Working on 33% space above the fire bars to 66% below should provide enough clearance.

BELOW LEFT: Loughborough University School of the Arts Anagama wood-fired kiln. *Photo: Author*

BELOW RIGHT: Anagama firebox. Air is drawn into the kiln through the four channels which run under the grate. *Photo: Author*

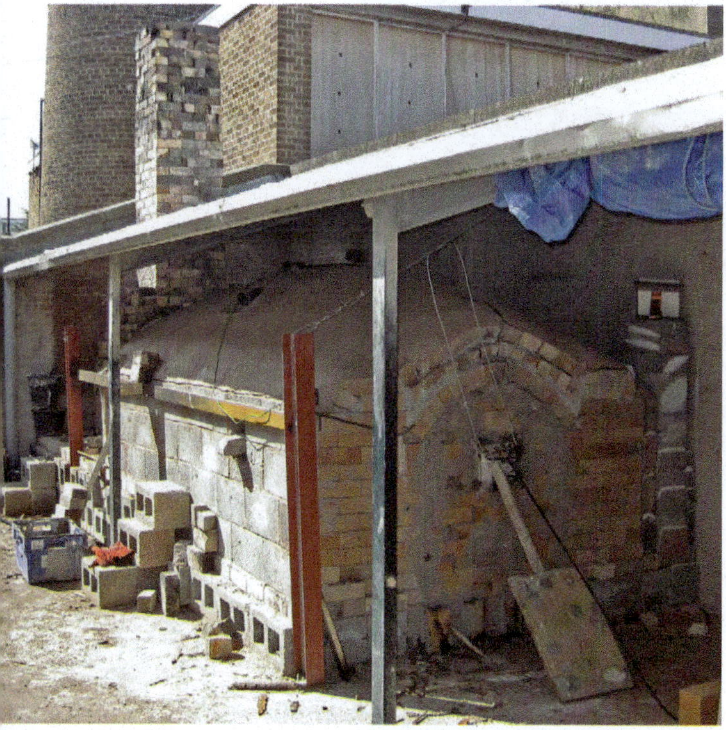

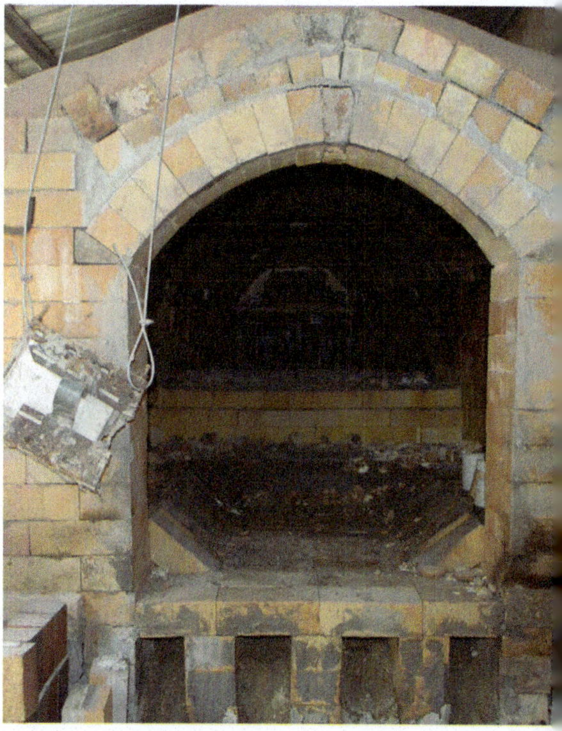

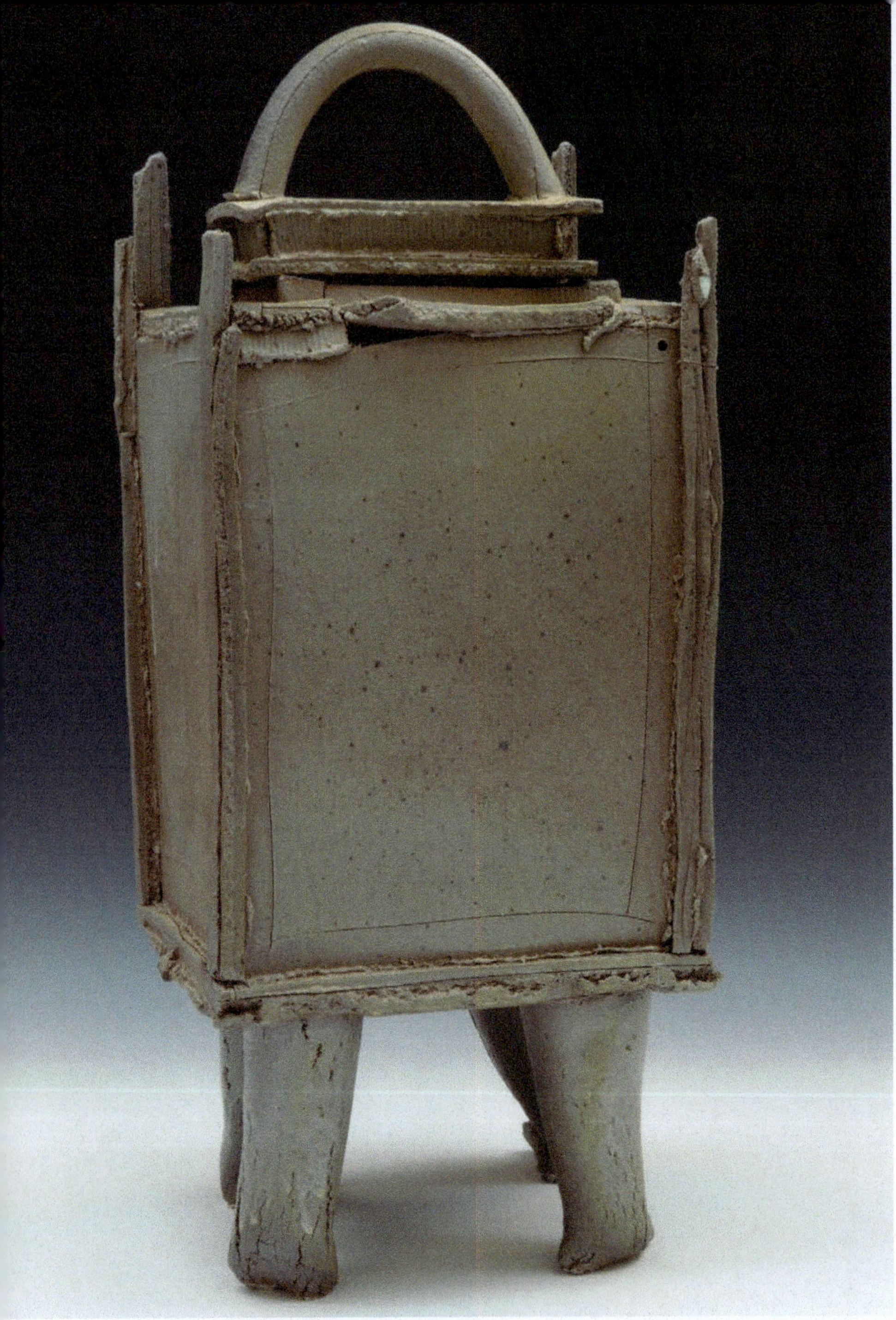

LEFT: *Box*, Benedict Brierley. Anagama wood-fired for three days to 1300°C (2372°F), 50 x 30 cm (19½ x 12 in). *Photo: Author*

A consideration when planning this type of kiln is fuel storage. Wood is a bulky fuel, and does require a significant amount of space which obviously must be kept dry. It will also probably need to be processed in some way; this usually involves slab wood being cut to a particular size to fit specific kiln fireboxes. Fuel can then be stacked in a convenient place for introduction into the kiln.

View through one of the fireboxes in Stephen Parry's Phoenix-style crossdraft wood-fired kiln. At the end of the firebox is the opening into the chamber. Air is drawn in under the fire bars and preheated by the ember bed. *Photo: Author*

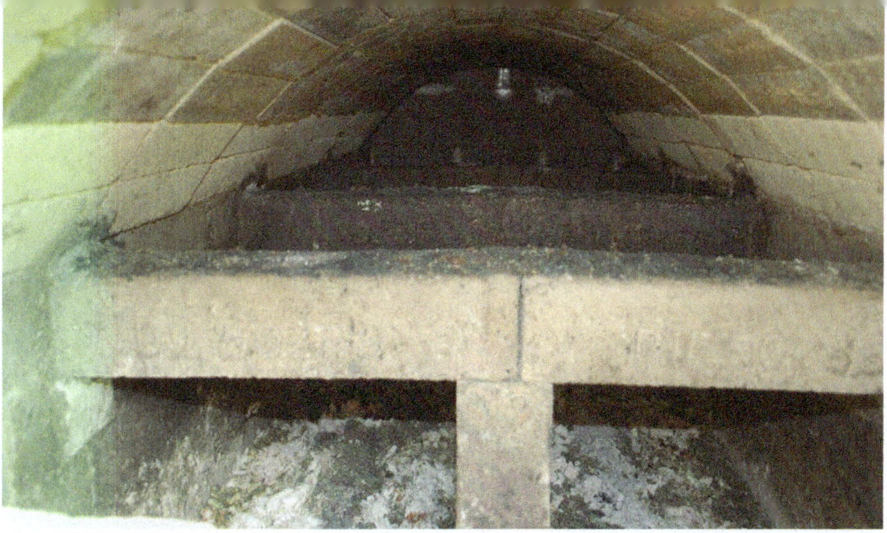

Small square earthenware plate, Patia Davis. Wood fired in her updraft bottle kiln.

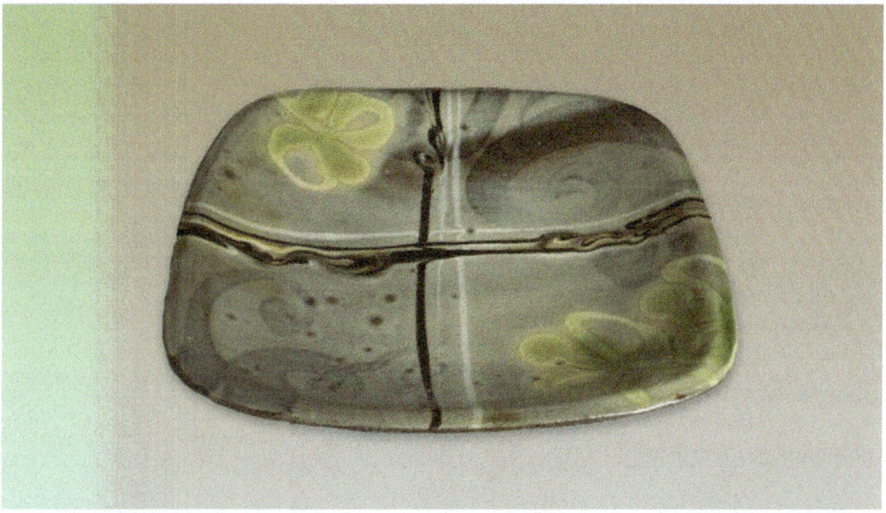

Wood-kiln fuel drying under cover. *Photo: Author*

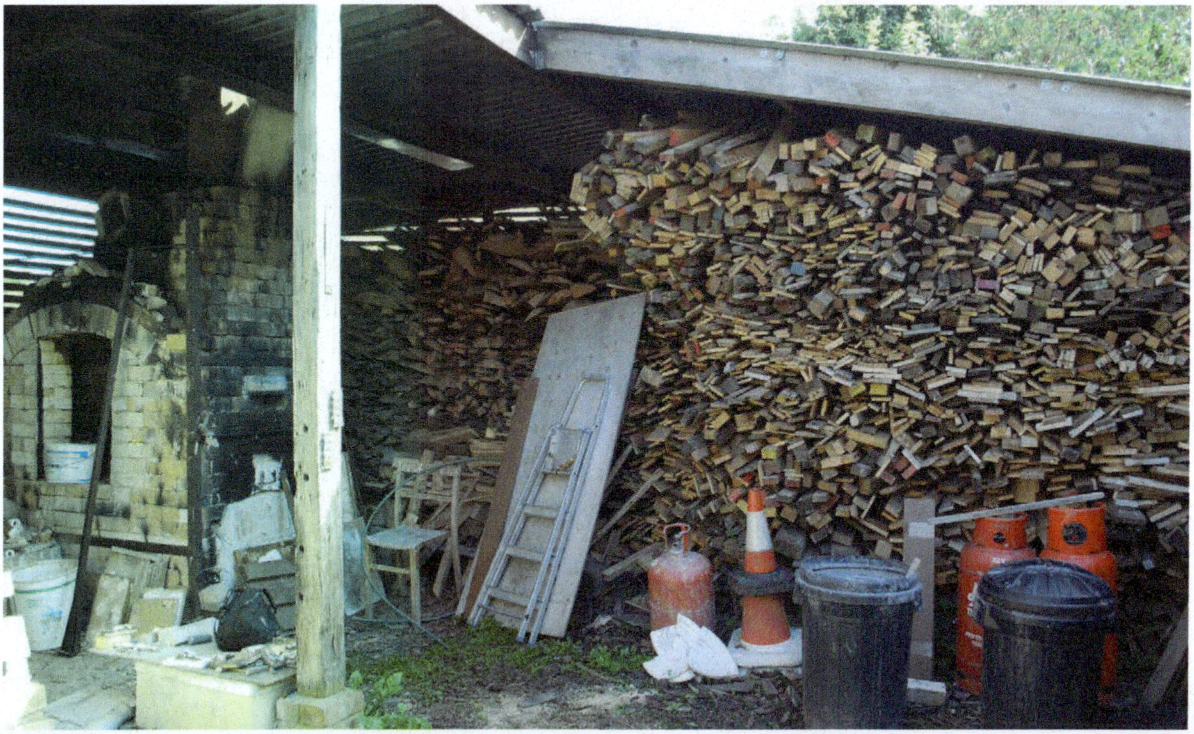

Oil

Oil-fired kilns can be of updraft, downdraft or crossdraft design. Oil used for firing can also vary; heating oil is a common form, but supplies such as old engine oil or even vegetable oil could also be used. Heating oil itself is not a highly flammable fuel: if one dropped a match into a pool of heating oil the match would be extinguished. As this fuel is not so readily flammable, there are several parts which make up an oil-burner system, and there are several differing designs.

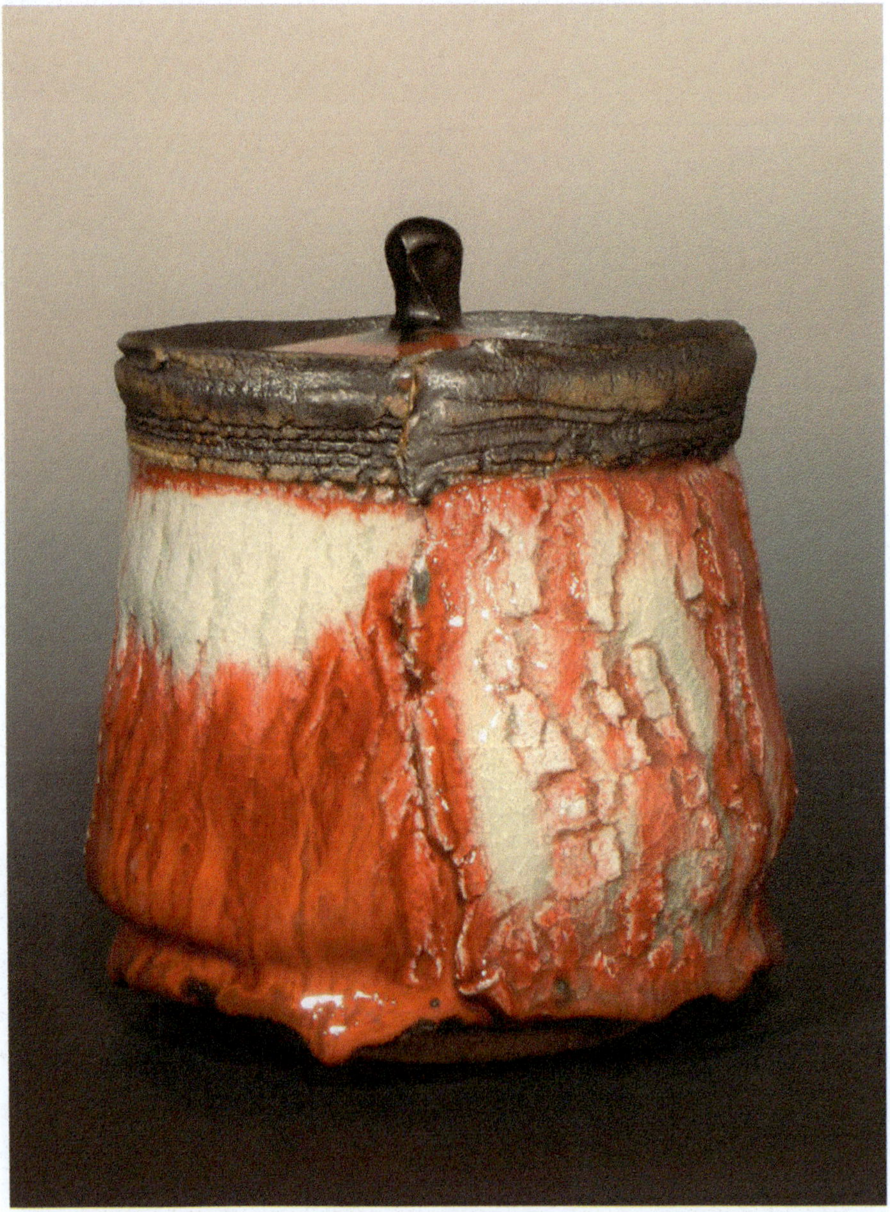

Jar, Eddie Curtis. Oil-fired copper glaze in reduction, 16 x 12 cm (6½ x 4¾ in). *Photo: Eddie Curtis*

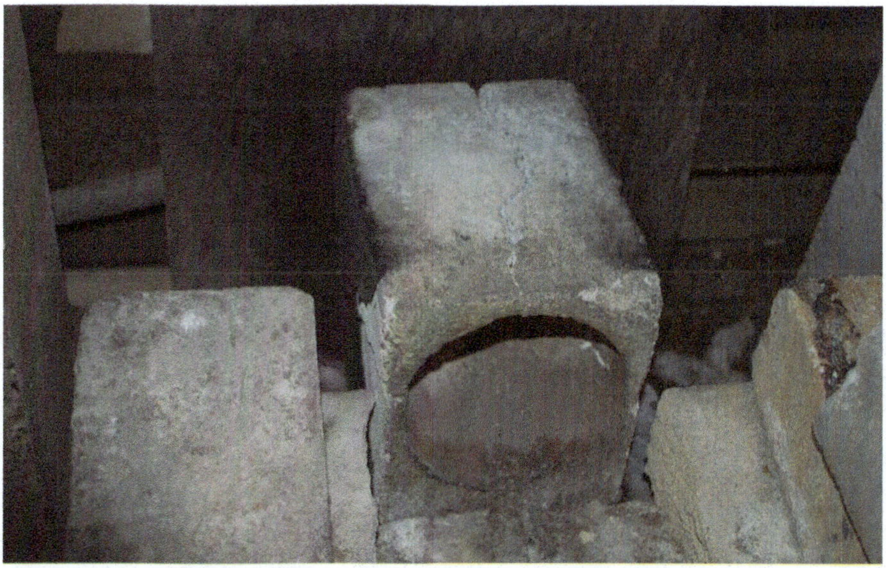

Quarl. This block made from refractory castable is preheated with either wood or gas to a temperature sufficient for the oil to combust on contact when sprayed from the atomiser. *Photo: Author*

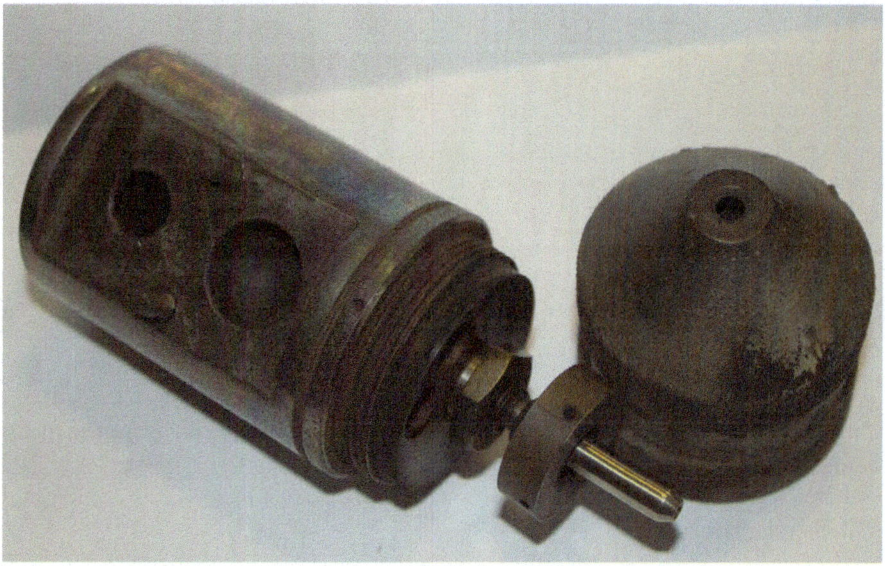

Internal workings of a 'swirlamiser' atomiser. Gravity-fed oil passes down the central 'needle'. Compressed air is forced through the small holes drilled into a brass disc and then through the end cap, around the oil in a swirling fashion. Hence, the term swirlamiser. *Photo: Eddie Curtis*

Atomiser and quarl

This system uses a two-part burner and burner block known as a quarl. The quarl is a block, usually made using refractory castable or firebrick and has a tapered aperture which runs through it, the wider end of which faces into the firebox of the kiln while the narrower end faces out. On the outside of the kiln is the atomiser, connected to an oil feed and a forced-air supply, usually a fan.

In order for the system to work, the quarl has to be heated to a point where it is hot enough for the oil to combust on contact when sprayed onto it. Gas or wood are good initial sources of heat to warm the quarl and once it is hot enough, the oil and air mix

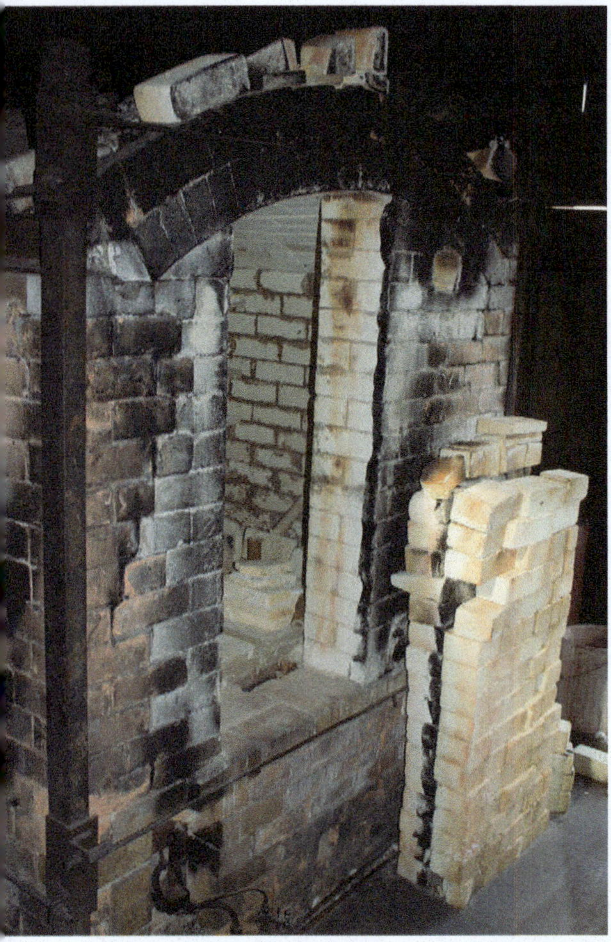

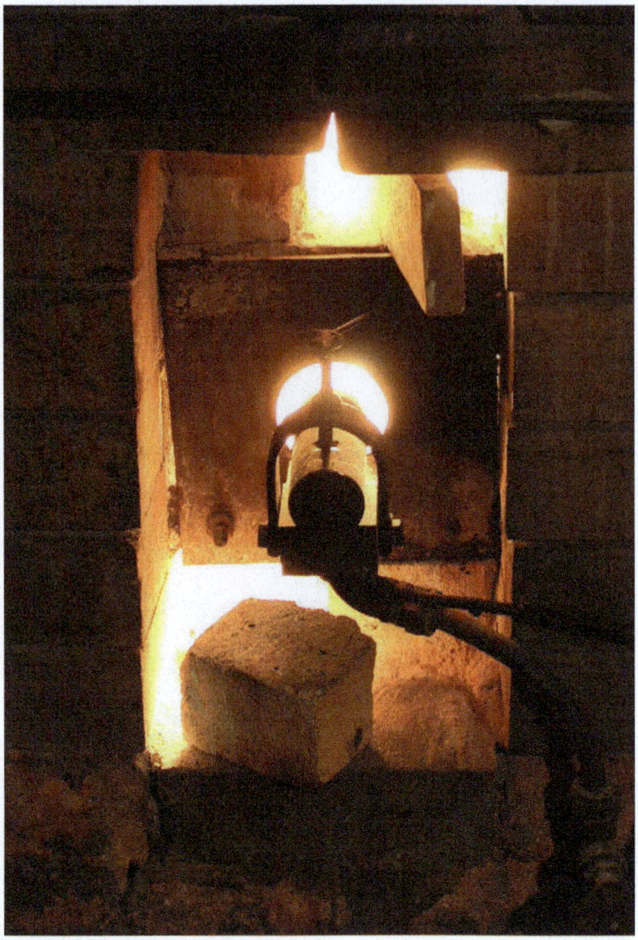

is turned on and the oil ignites within the quarl, forming the flame which extends into the firebox. Oil can be a very effective fuel which allows for very good control of kiln temperature and atmosphere.

ABOVE, LEFT: Eddie Curtis's oil-fired downdraft kiln. *Photo: Eddie Curtis*

ABOVE RIGHT: One of Eddie Curtis's 'swirlamiser' oil burners firing. *Photo: Eddie Curtis*

Oil-drip burners

Drip burners do not necessarily require a fan to introduce the air into the combustion mix, although a forced-air supply gives more control options. They consist of a series of overlapping metal plates mounted in a vertical frame and placed in front of the fire mouth, these plates are preheated either with gas or wood as with the quarls above.

Oil is gravity fed from a header tank on to the top plate and flows down over the others. The oil ignites on contact with the hot plates and forms the burner. Air is drawn or forced through gaps between the plates and provides the fuel to air mix. Greater heat generation can be achieved by increasing the flow of oil. If a fan is not to be used then the height of the chimney would dictate the amount and speed at which air will be pulled through the drip plates, in this case, a damper in the chimney could be used to control the flame length and therefore temperature climb.

RIGHT: *Serving tray,* John Calver. Fired in reduction to 1320°C (2408°F) in his oil-fired kiln, length 45 cm (17¾ in). *Photo: John Calver*

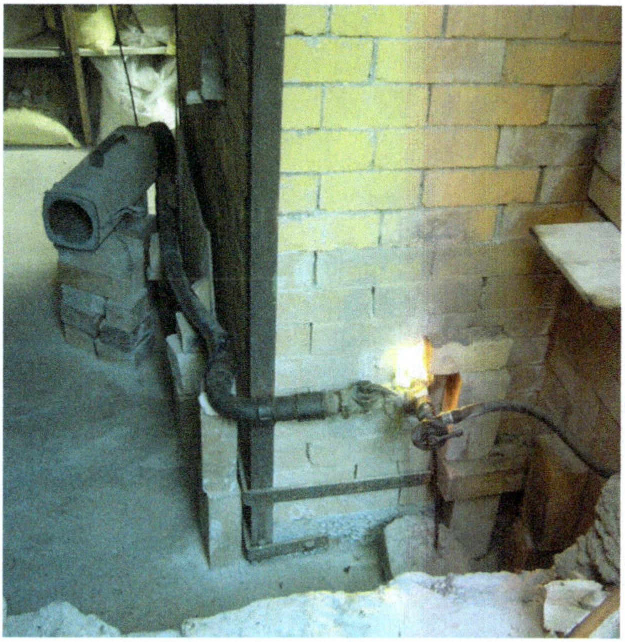

Oil-burner firing in John Calver's oil-fired kiln. *Photo: John Calver*

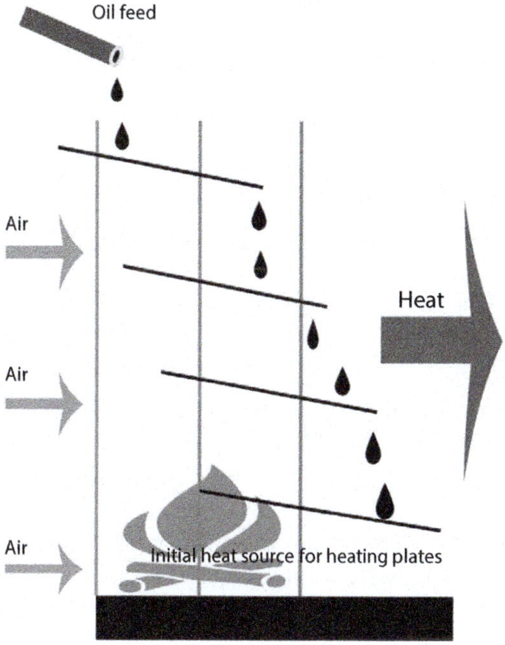

Diagram: Oil-drip burner.

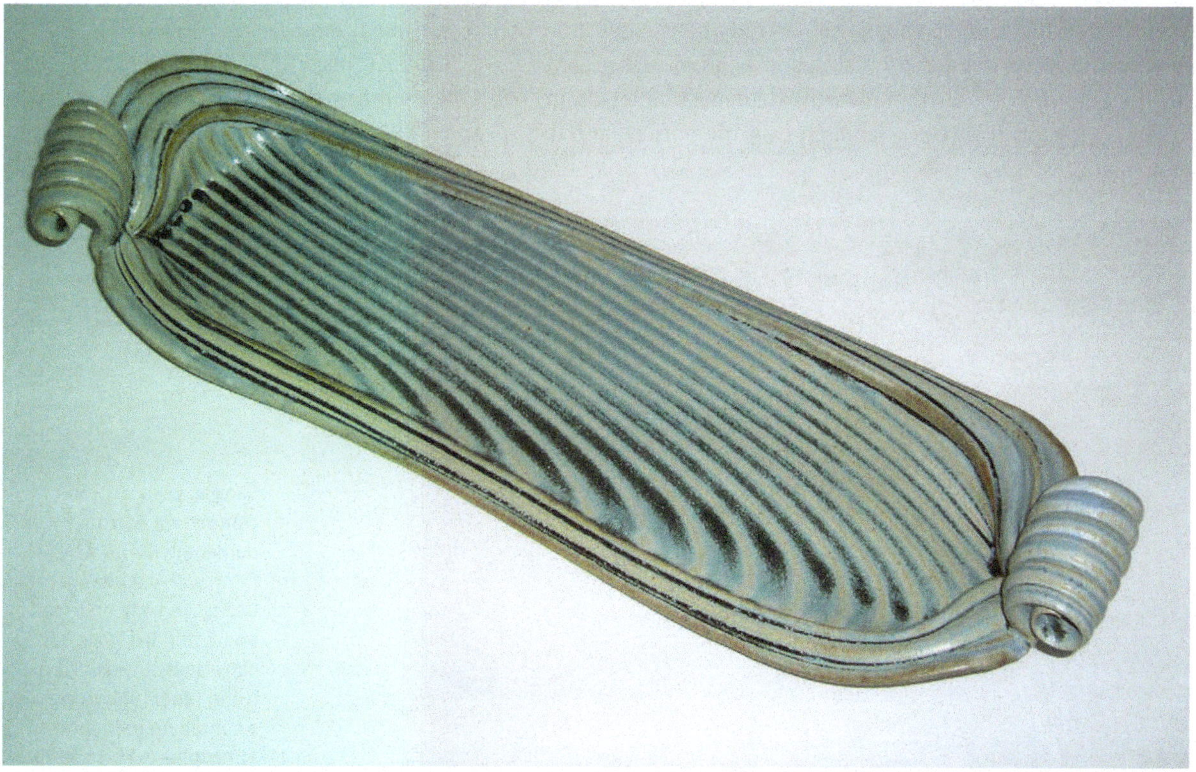

Coal

Coal as a kiln-firing fuel has pretty much ended in Europe and the USA. The biggest example of the historical use of this fuel is the industrial archaeology of Stoke-on-Trent in the UK. One of the reasons for the emergence of this area for large-scale production of pottery was due to the abundance of coal, lying as it does on the north Staffordshire coal seam.

One of the problems which had to be overcome with using coal as fuel was the amount of sulphur released when it was burnt. This sulphur reacted very badly with the glazed wares. For this reason most of the pottery fired in these kilns was fired in saggars – a saggar is essentially a box built from fire clay in which is placed the wares to be fired, thus protecting them from the adverse kiln atmosphere.

Coal-fired kilns tended to be updraft in design, mostly due to coal producing a short flame. Fire mouths would be positioned around the outside of the chamber with the heat produced being deflected under the packing space where it would be drawn up through the wares. Firing work in saggars also provided a convenient way of stacking work in the chamber by placing saggars on top of each other in towers.

Other fuels

Other kiln fuels which have been explored include paper, whereby the whole kiln is built out of densely layered paper. These kilns don't reach high enough temperatures to instigate the ceramic change so can only be used as a decorative surface firing. In some parts of the world brush wood and old vehicle tyres have also been used to fire kilns. Indeed, anything which is combustible could in theory be used to fire a kiln as long as it produces enough calorific output, and it is given enough air to combust efficiently.

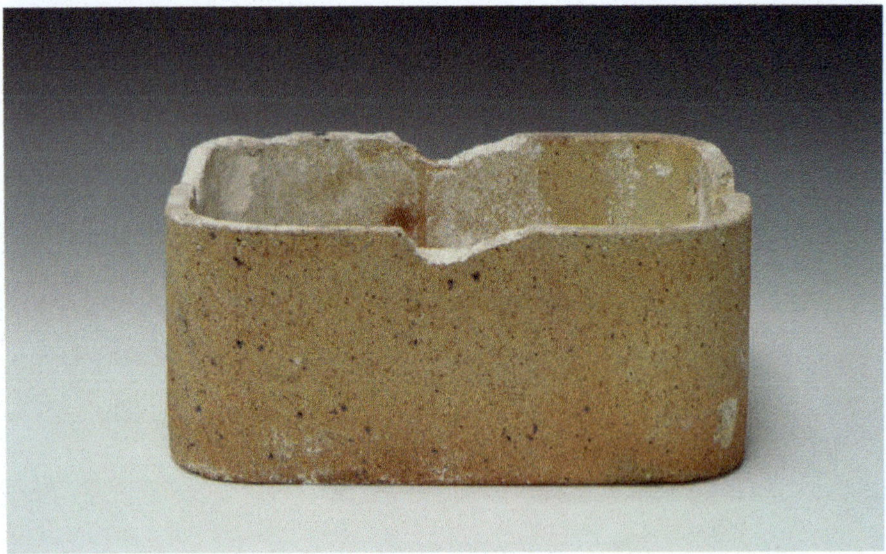

Saggar: used to protect ceramic wares from damaging fumes when firing with coal. Saggars can be stacked one on top of another in tall stacks. They can also be used in kilns using other fuels to create a contained reduction atmosphere. *Photo: Author*

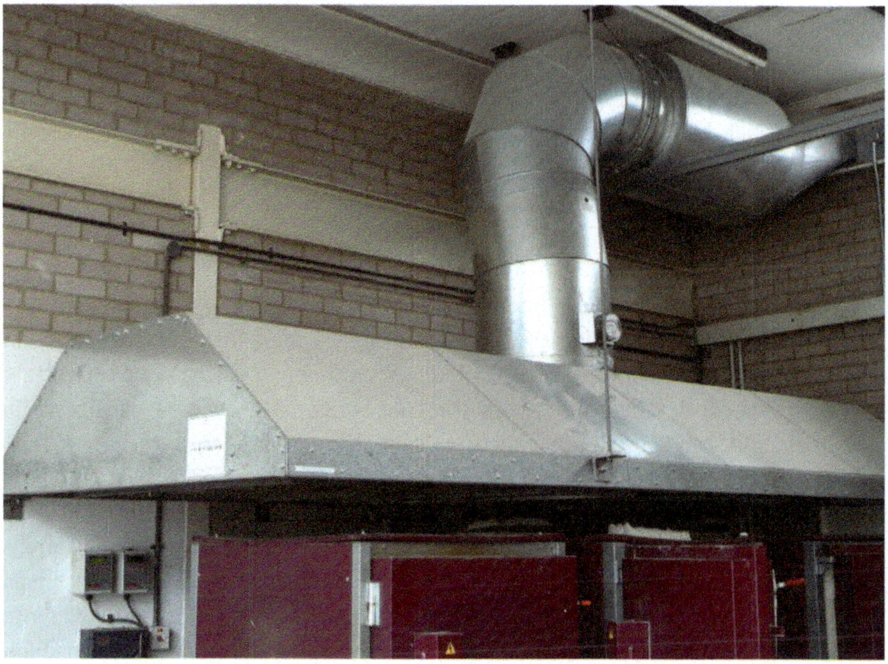

The large extraction hood covering three electric kilns at Loughborough University School of the Arts. *Photo: Author*

SETTING UP A KILN

As we have seen above, there are many different designs of kilns, all of which need careful consideration when planning where to situate a kiln and deciding how appropriate they are for the job in hand. If a maker is interested in low-temperature raku work for example, there is little point building a 150 cu. ft anagama kiln. Once a choice of kiln has been decided upon, the details of where the kiln should be sited can be addressed. Wherever a kiln is placed, it must be well ventilated and allow for the safe movement of work around it.

For an electric kiln an outhouse with a window will suffice, as long as the structure is waterproof and allows for enough clearance around and above the kiln. However, if it is a fuel-burning kiln it needs to be well vented via an external chimney, as well as having plenty of natural ventilation around the kiln. Kilns should be sited on a solid and sound floor with plenty of space overhead; for an electric kiln it is recommended that a space of at least 30 cm (17½ in) be left between the kiln walls and the building walls.

It is also worth considering how the kiln will be accessed. If it is a front-loader kiln, it is preferable to have good clearance for the door to open so that it is not in the way when packing and unpacking work. A top-loading kiln does not require this extra work space so is possibly a better option in workshops with limited space. The potential downside of a top-loader is that they involve a lot of bending to lower shelves and work into the chamber.

LOADING KILNS

When packing work into any kiln, there are several generic points which should be observed. Each kiln, dependent on design and fuel, will then have its own specific requirements.

Kiln furniture

Kiln shelves or batts are the refractory shelves on which the work is placed; these range in size, shape and thickness. They are generally made of either cordierite, an alumino silicate material, or silicon carbide. Silicon carbide shelves are lighter and stronger than conventional cordierite shelves but are significantly more expensive. The most common shelves are those made from cordierite.

Kiln shelves are supported using kiln props; again these are made from cordierite. Props come in many varying lengths and diameters. Some are castellated to facilitate locking, some are simply tubes. Any refractory material can be used as shelf props; for example in the anagama kiln that I fire, refractory kiln bricks are used which have been cut in half lengthways. In a kiln that one is throwing large pieces of wood into, the extra stability offered by solid fire brick props helps to make everything good and secure and less likely to be unbalanced by a careless stoke.

Refractory components can be obtained to perform many bespoke tasks such as tile cranks and plate racks. Not only do these refractories help to overcome some of the issues involved with supporting flat ware in the kiln, but they also save space allowing more work to be packed.

Loading a kiln may be approached by packing the floor first and then moving up a level, or from the back of the kiln forward. Whichever approach is taken is really down to personal preference as well as being dictated by the type and shape of work to be fired, and the type and shape of kiln.

FACING PAGE

TOP ROW, LEFT TO RIGHT: Plate crank. Used for stacking multiple plates efficiently when firing.

Tile crank. Used for stacking smaller plates or tiles efficiently when firing.

BOTTOM ROW, LEFT TO RIGHT: Three-point shelf propping. This style of shelf support spreads the load across the edges and middle.

Another three-point propping technique.

Four-point shelf propping.

Photos: Author

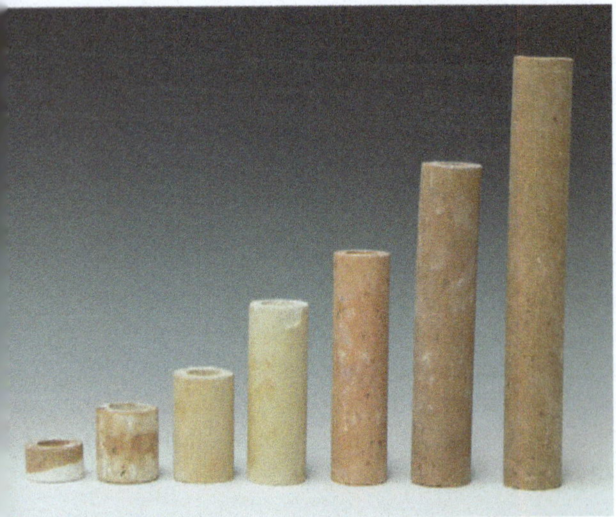

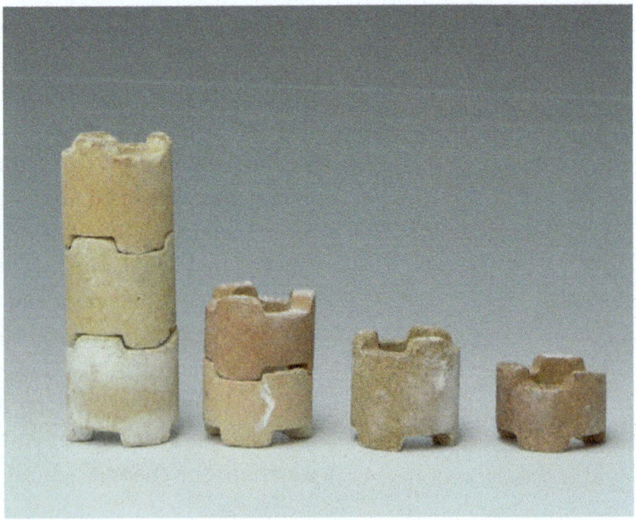

BELOW LEFT: Range of tube kiln shelf props.

BELOW RIGHT: Range of castellated kiln shelf props.

Photos: Author

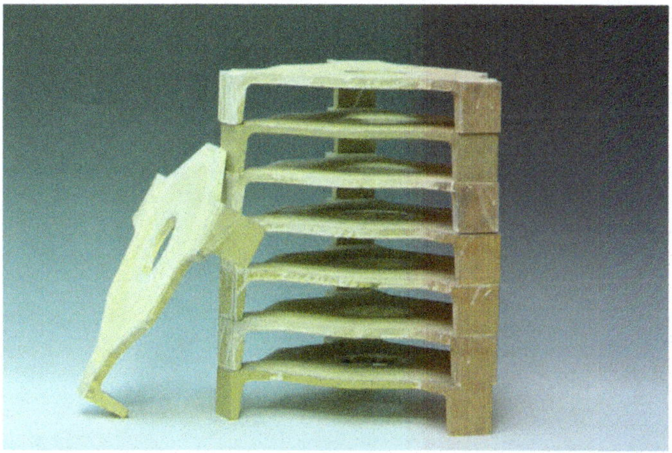

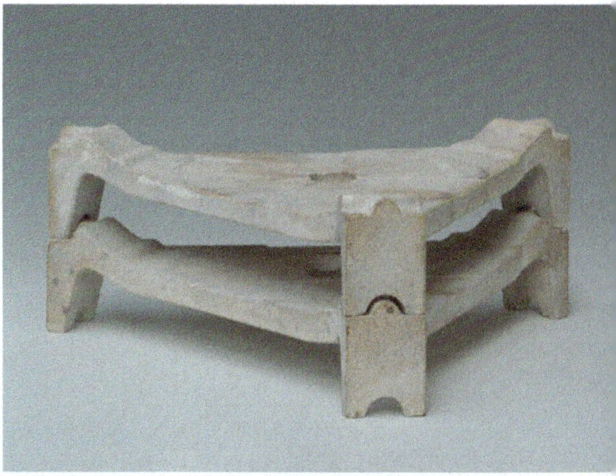

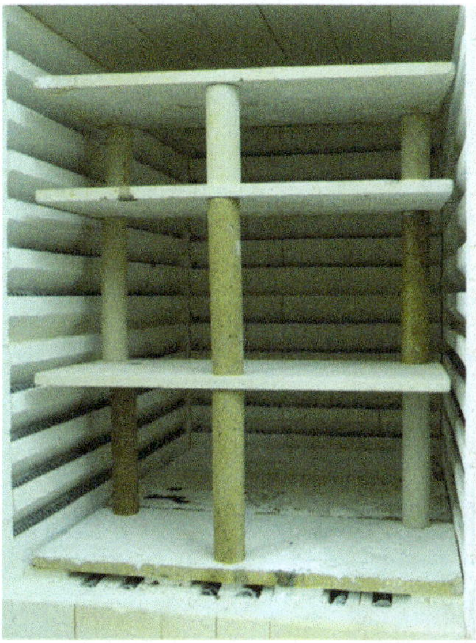

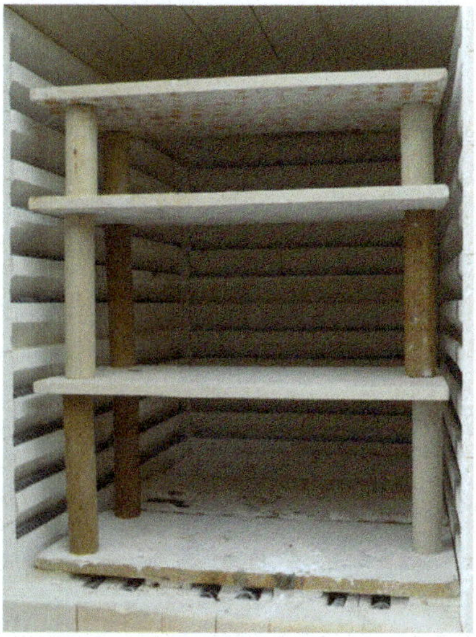

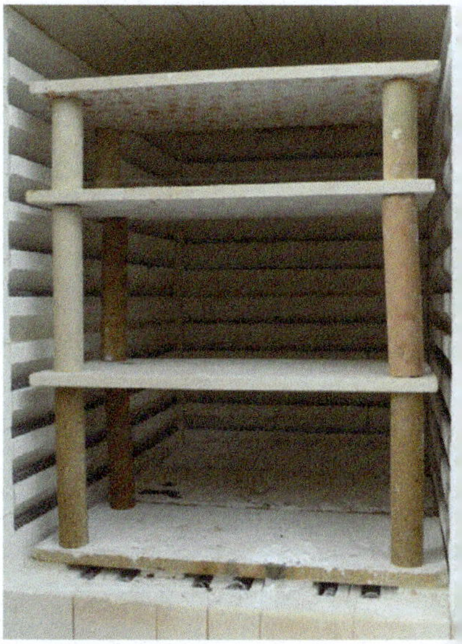

There are, however, several points which can promote a strong and stable pack and reduce the risk of a shelf failure during firing. Generally a three-prop supporting system is the most stable; this can be either two at the rear of the shelf and one at the front or vice versa. By using this method, the shelf is supported both at the edges and in the centre. This is equally appropriate for a circular-shelved top-loading kiln.

There may be instances where it is not possible to support the shelves at three equidistant points, in which case, four props should be used at the corners of the shelves. Once the first props have been positioned it is important that as much as possible, the props continue up the kiln in straight lines so that the first prop supports the first shelf and the next prop sits directly above, and so on. The reason for this is so that the majority of the load is transferred through the props to the kiln floor and once again avoids the shelves taking too much load.

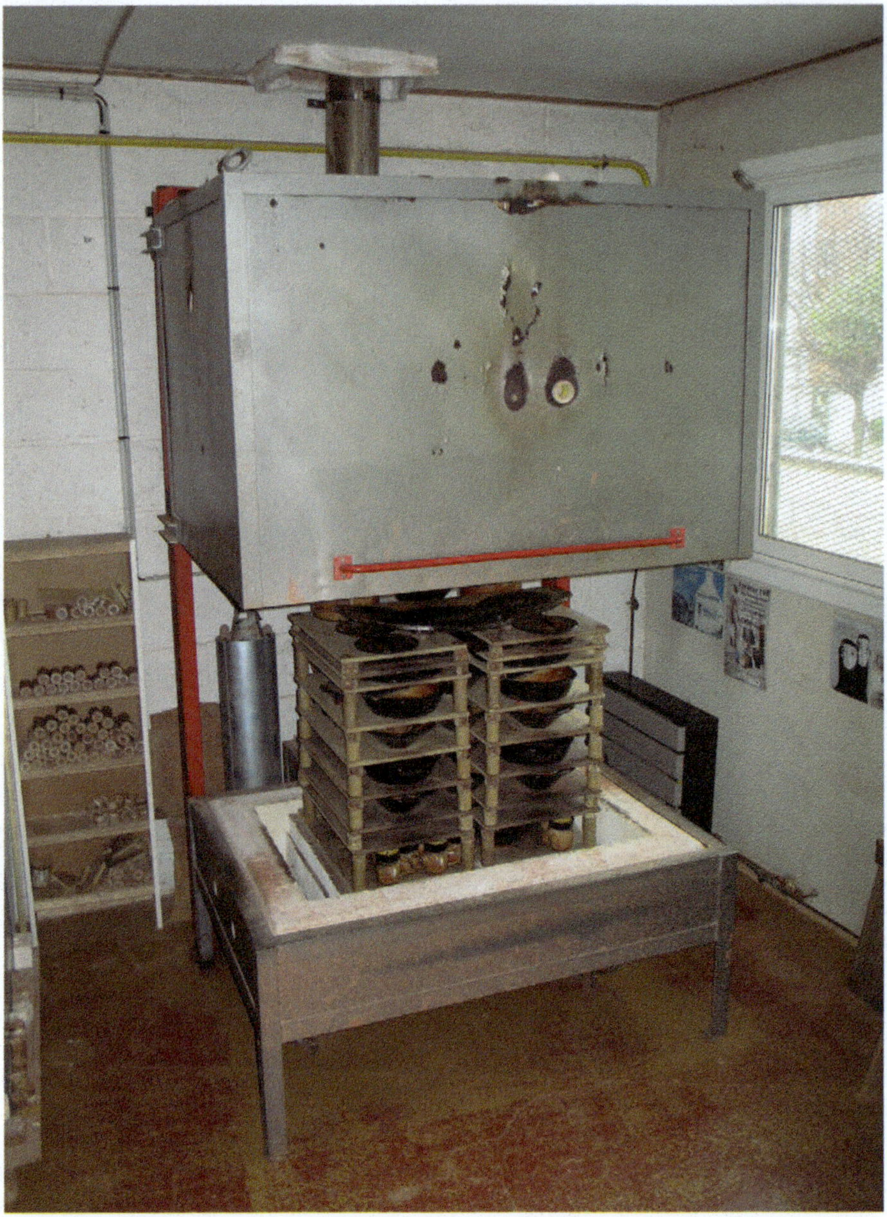

Françoise Dufayard's gas kiln stacked ready for unpacking. Françoise has used four-point stacking to accommodate larger flat ware. *Photo: Francoise Dufayard*

Other refractories which may be used when packing kilns are stilts or spurs. These are refractories which are designed to support work that has been glazed all over. These refractories come in all shapes and sizes and have pointed tips which, after the glaze firing, can be easily snapped off leaving only a small scar on the glazed surface. Generally this type of support is only used in earthenware firings because in stoneware firings the pyro-plasticity of the clay body can cause the points of the stilts to sink into the clay body making them too difficult to detach after the firing.

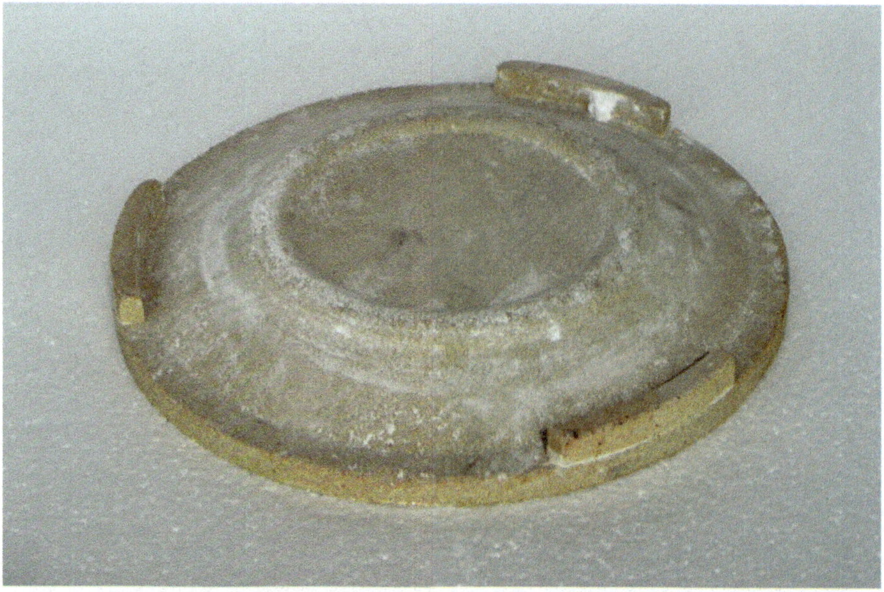

Small plate/bowl setter. Used for minimising distortion on rims. Pieces are placed upside down on the setter. The rim can then slide up the setter slightly as it shrinks. *Photo: Author*

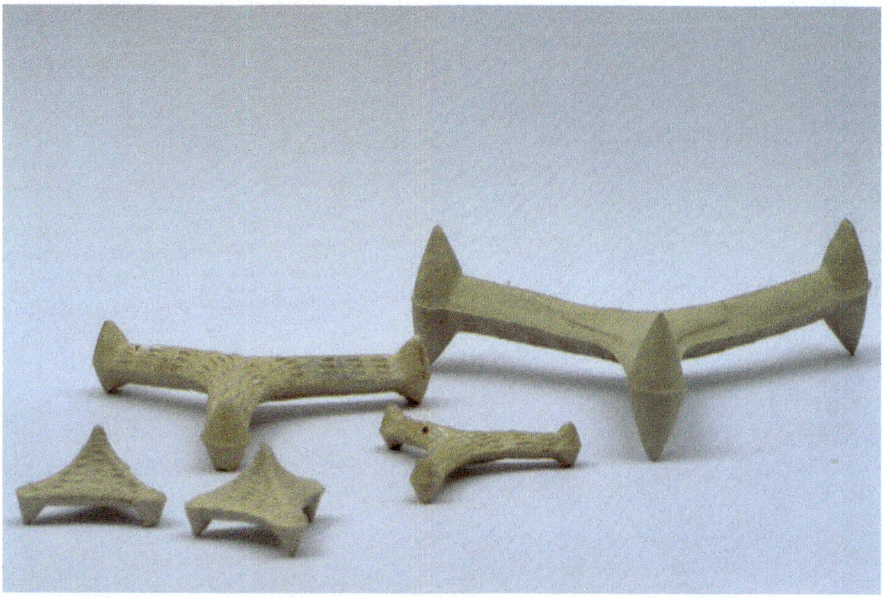

Refractory stilts for supporting work which is glazed all over including base. *Photo: Author*

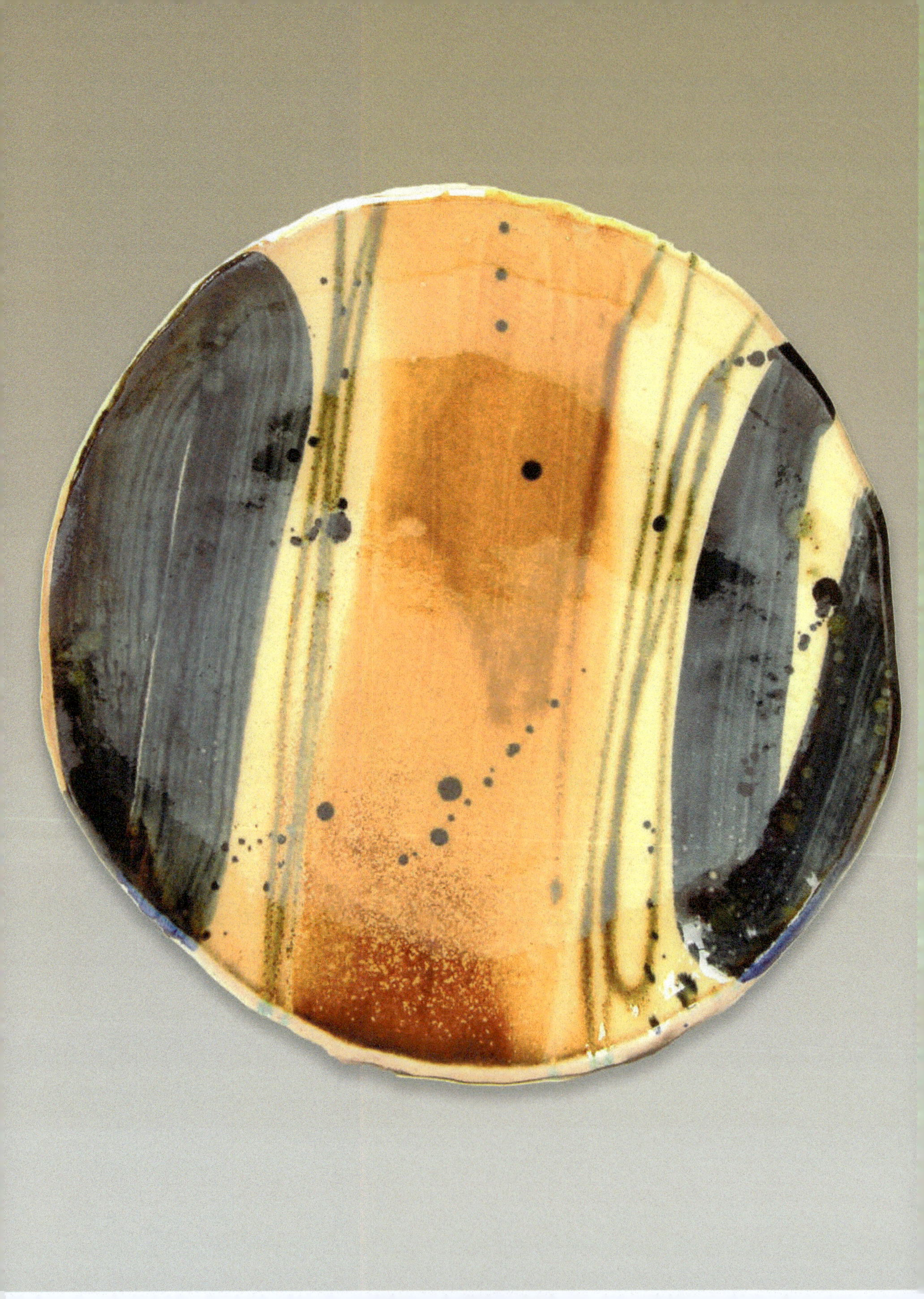

3

Firing in oxidation

Oxidised firing has become a blanket term which can describe two slightly differing kiln atmospheres: oxidised and neutral firings. In a neutral firing it could be said that there is sufficient oxygen available for the fuel to burn efficiently; in an oxidised firing there would be an abundance of oxygen. In reality it may be argued that achieving a balance by which the chamber atmosphere could accurately be described as truly neutral would be almost impossible to maintain throughout a firing so oxidised would probably be a more accurate description of this type of firing. A definition of an oxidised firing could therefore be: a firing in which there is plenty of oxygen available for the fuel to be combusted efficiently and completely.

OXIDATION: ELECTRIC KILNS

In an electric kiln, there is no fuel directly involved in the generation of temperature, and the internal chamber space is heated by radiant heat emitted by the elements which are housed in the walls, floor and sometimes the door of the kiln.

Although there is no fuel combustion within the kiln chamber there may be some oxygen removed from the kiln atmosphere by the combustion of organic material contained within the clay, which will be dependent on the type of clay being used. Some highly processed clays will be very low in organic matter, whereas some less refined clays may be quite high.

The author, however, has experienced a slight discolouration of clay bodies due to accidental reduction in an electric kiln containing some work made from paper clay in its biscuit firing. This was in a relatively new electric kiln with no spy holes through which oxygen could be drawn to replace that used by the combustion of the paper fibres. The exit flue of the kiln had also been closed after the initial drying period of the firing and so carbon monoxide produced by the paper fibre combustion was contained within the kiln and was free to liberate oxygen from the ceramic oxides within the clay. This mild low-temperature reduction manifested itself in a slight darkening of the biscuit-fired body, along with a reduction in the porosity of the biscuit ware due to a slightly higher vitrification compared to the same body fired in a kiln in which there was no paper clay being fired.

Generally though, electric kilns fire with a neutral atmosphere. As mentioned above when firing there will be some gases emitted. Initially there will be water, followed by organic combustible materials remaining in the clay; next, chemically combined water is liberated and then other gases such as small amounts of sulphur. None of these stages will significantly alter the kiln atmosphere under normal circumstances.

LEFT: *Large hand built platter.* Nigel Lambert. Amber glaze and cobalt brush work. Raw glazed. Wood-fired earthenware in oxidation, diameter: 45 cm (17¾ in). *Photo: Nigel Lambert*

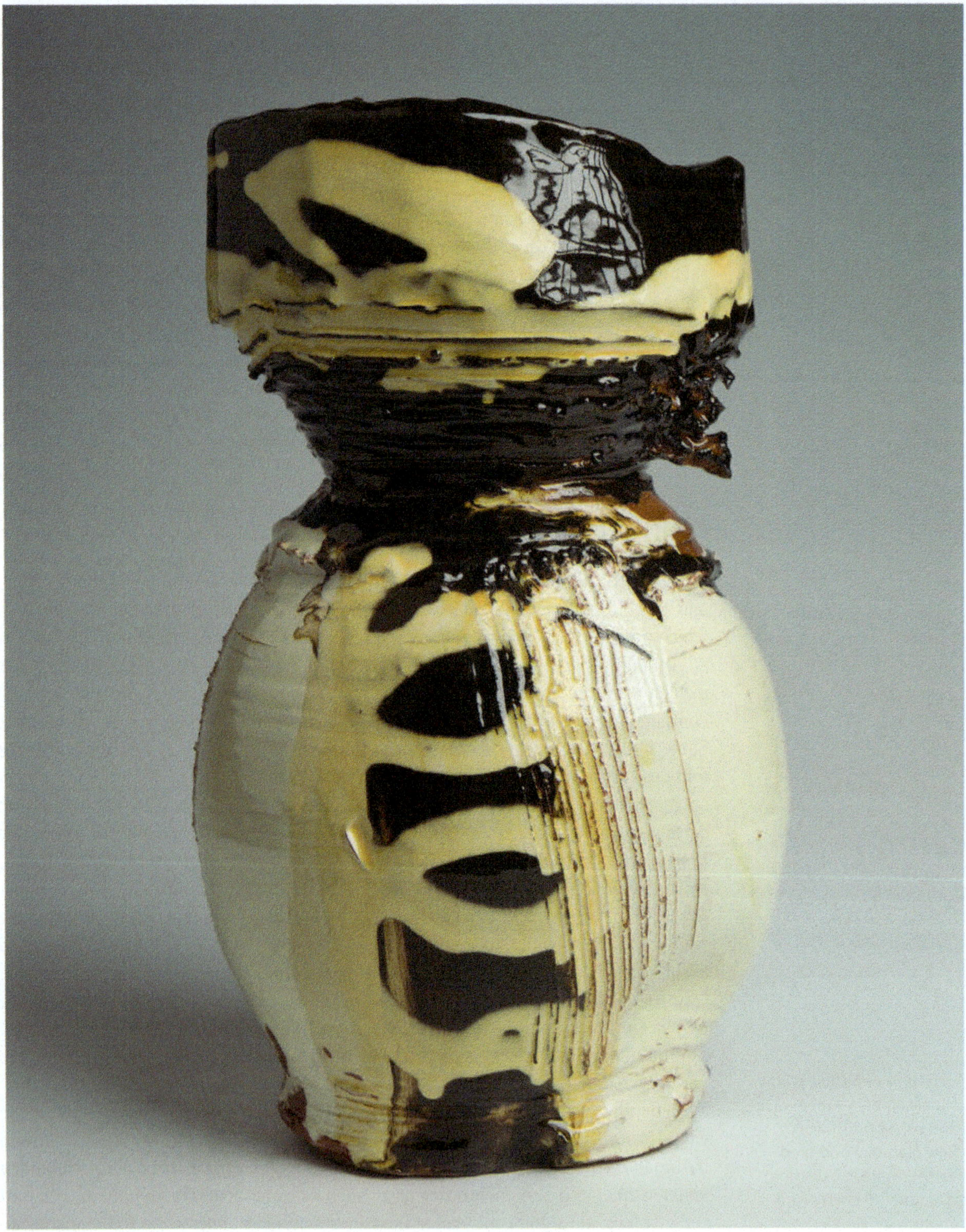

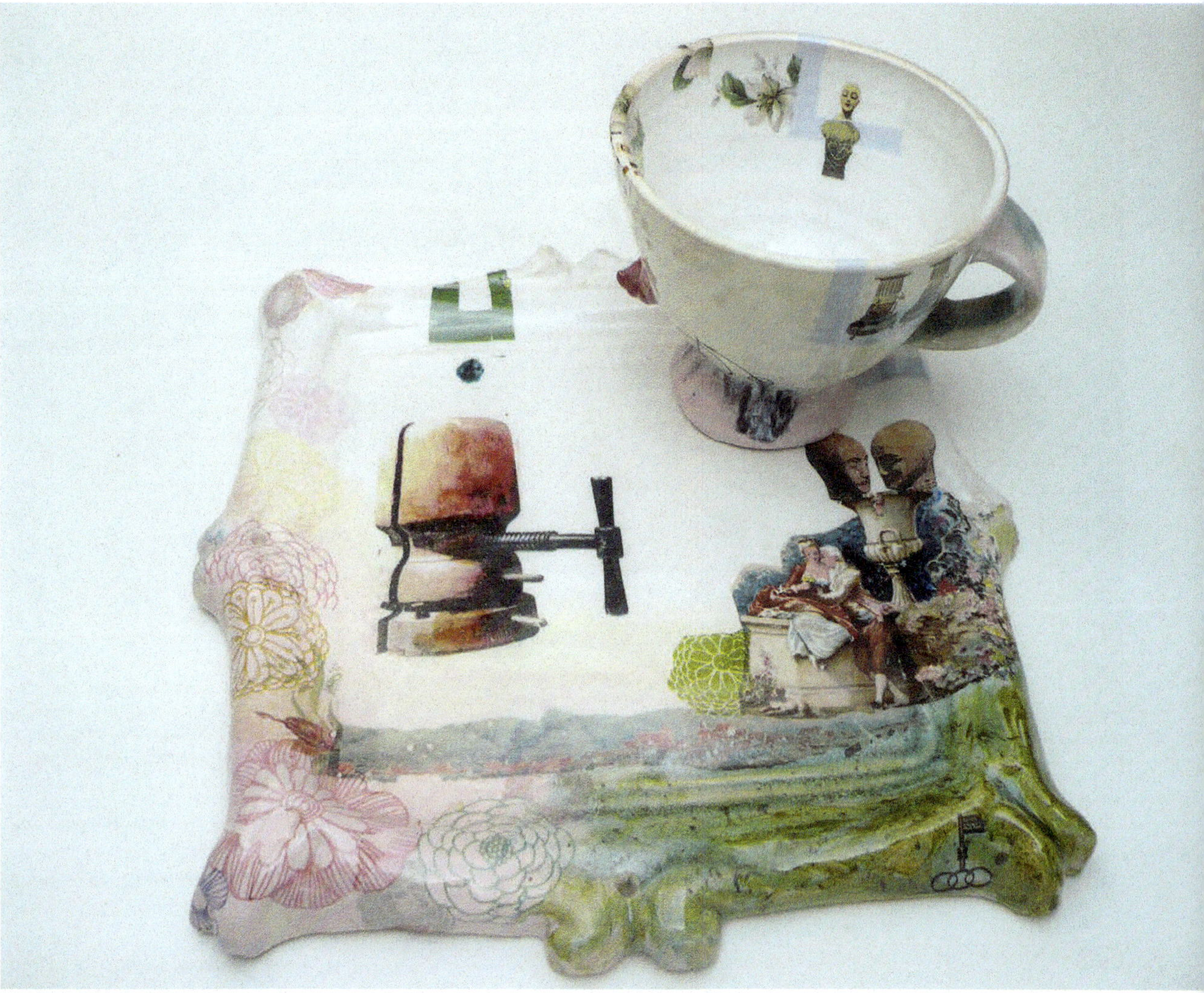

ABOVE: *Platte Vejle,* Lone Borgen. Electric kiln-fired to 1180°C (2156°F), followed by a transfer firing to 840°C (1544°F), also in her electric kiln, 30 x 30 cm (17½ x 17½ in). *Photo: Lone Borgen*

LEFT: *Thrown vessel,* Dylan Bowen. The work is fired in an electric kiln, to around 1060°C (1940°F) and then soaked for an hour or so at around 1050°C (1922°F), 32 x 24 cm (12½ x 9½ in). *Photo: Ben Ramos*

It is important, however, that even an electric kiln be vented or at the very least, be situated in a workshop which has plenty of natural ventilation to avoid any build-up of fumes in the working environment.

OXIDISED ATMOSPHERES IN FUEL-BURNING KILNS

When firing a fuel-burning kiln the process of firing in an oxidised atmosphere becomes a little more complex. As has been discussed earlier, a fuel-burning kiln relies on a balance of fuel and air to generate a temperature climb efficiently. Each kiln will have slight variations depending on the fuel being used, and the overall design of the kiln. We will have a closer look at some of these considerations.

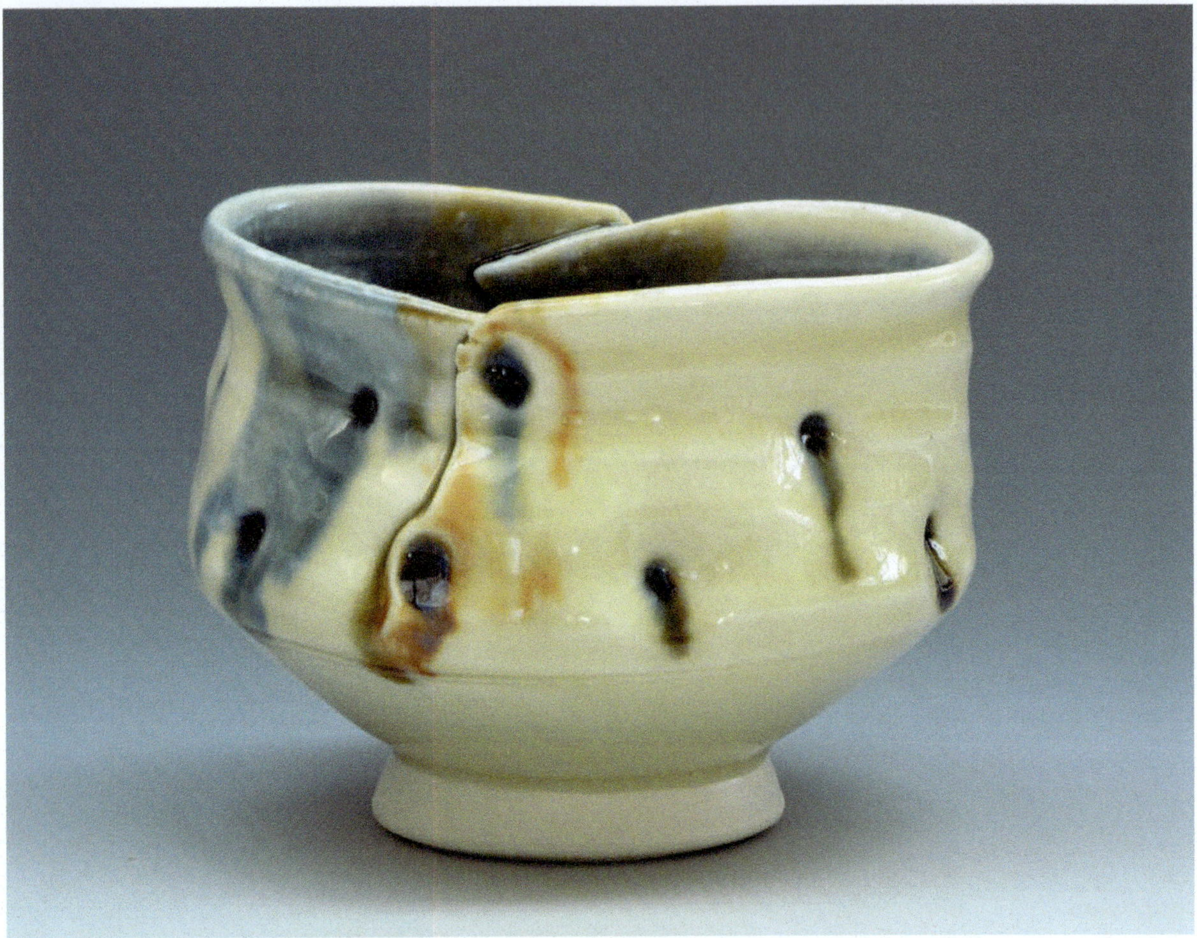

Teabowl, David Scott. Fired in an electric kiln to 1240°C (2264°F), 8 x 11 cm (3 x 4½ in). *Photo: Author*

Gas

Gas kilns generally have two main ways of introducing oxygen into the combustion mix; natural and forced air. In order to achieve an oxidised atmosphere in a naturally aspirated kiln a balance has to be found between the fuel and the air. To add to the equation, gas kilns may be fitted with different-sized burners. For example, a naturally aspirated kiln may have burners which when turned on full are capable of supplying more fuel than air is able to be drawn in. The primary air supply will be drawn in around the burners with the potential additional feature of other secondary air inlets; however, these are not always present.

The fuel in this instance, if the burners were to be turned on fully, would be unable to combust efficiently due to lack of oxygen, and in this scenario the kiln would naturally move into reduction. In order to fire in an oxidised atmosphere a balance would need to be found between the amounts of fuel being introduced through the burners, relative to the amount of air available; this would normally be by simply lowering the amount of fuel being introduced. The fuel–air mix may be judged by

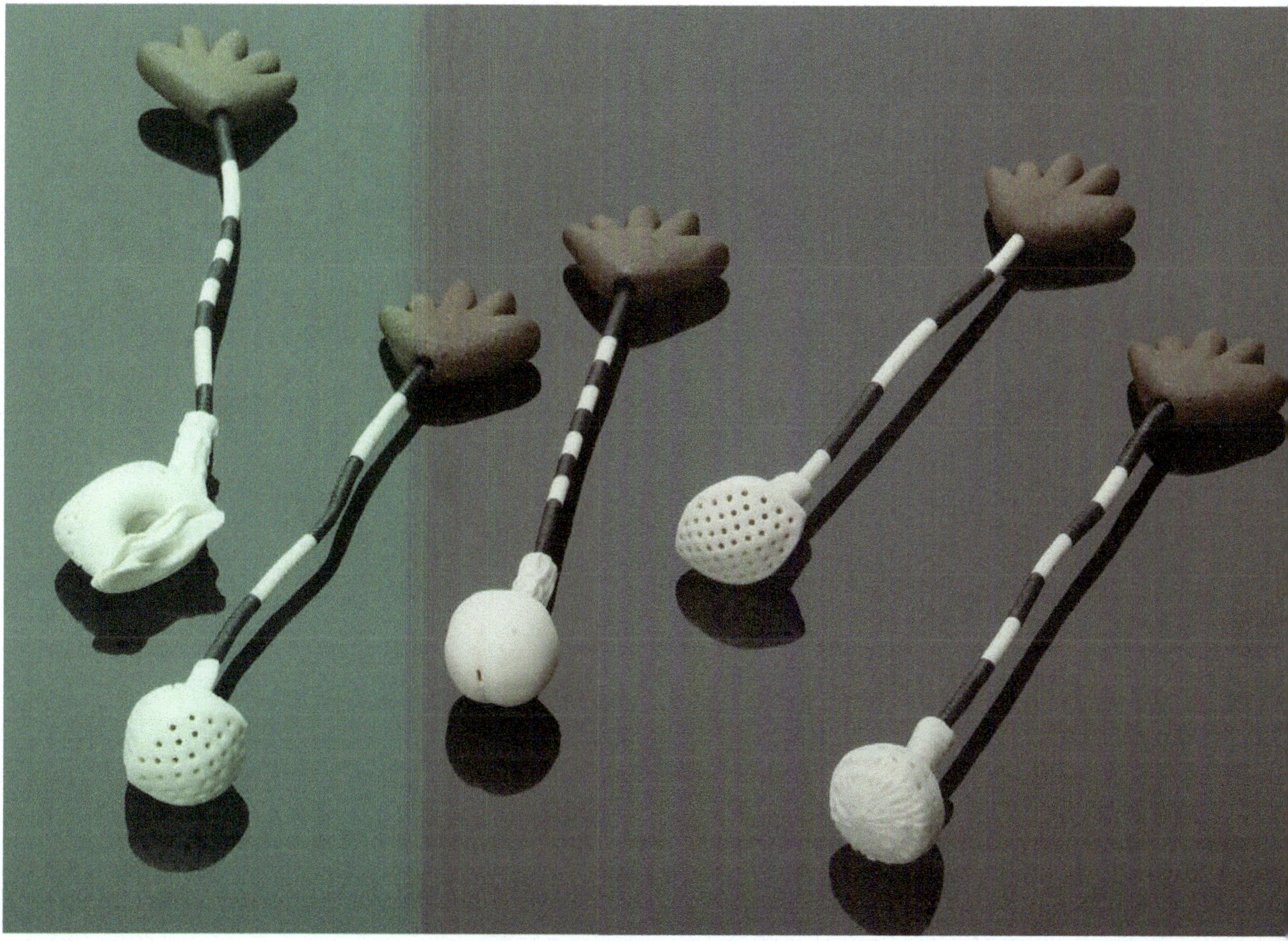

Shaken, Strained & Measured, Helen Marton. Gabbroic clay and porcelain fired in her electric kiln to 1260°C (2300°F), 25–30 x 4 cm (10–17½ x 1½ in). *Photo: James Mann*

observing the spy hole of the kiln. If the kiln is reducing then a small flame would be visible from the spy hole. The fuel level should be lowered until no flame is visible; the kiln could then be said to be firing in oxidation.

Obviously, enough fuel still needs to be added to achieve temperature climb, but not so much as to create a reducing atmosphere. A second scenario could be a kiln which is fitted with smaller burners or larger primary air inlets around the burners, a combination which should allow enough fuel to enter the chamber when on full to combust cleanly with the amount of oxygen available. In this case the balance is somewhat easier to find.

With naturally aspirated burners the primary air source would be around the actual burner. Secondary air may be introduced into the kiln by way of air inlets in the bottom of the kiln's side walls. The secondary air inlets introduce oxygen into the path of the flame allowing it to burn more efficiently. Most gas kilns will also have a damper in the chimney; this may be made of refractory kiln shelf or it may be an HTI kiln brick. Its purpose is to restrict the exit of gases and hold flame and heat within

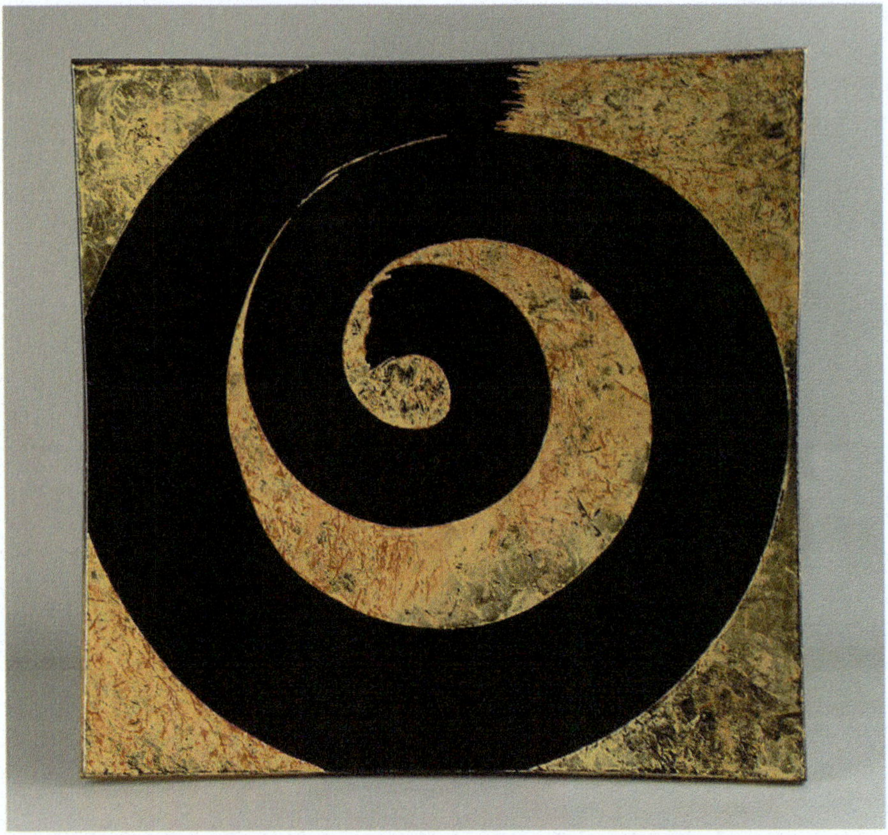

The Still Point, Françoise Dufayard. Oxidised earthenware, 65 x 65 cm (25½ x 25½ in). Françoise fires in a top hat gas kiln. She lights one burner initially, then another one 1 hour later until the four burners are operating. The gas is then increased three times (after each hour). The kiln temperature rises around 100C°/hour (212°F/hour). *Photo: Jean Blanchais*

the chamber; it also slows the amount of air being drawn into the chamber through both the primary and secondary intakes, therefore altering the fuel–air balance. In order to help achieve an oxidised atmosphere the damper would normally be left open allowing as much air as possible into the chamber. It is, however, possible – if the kiln exit flue is large – for too much heat to be lost out of the flue and temperature climb to be negatively affected. This scenario would require adjustment of the damper to hold enough heat in the chamber without putting the kiln into reduction.

With a gas kiln fitted with a forced-air system one has the added control of being able to physically increase or reduce air entering the fuel–air mix, by adjusting the air valve. With some systems, however, the fan delivers a fixed quantity of air and it is only the gas and damper which can be adjusted. In this situation, fuel in the early stages of a firing would gradually be increased to raise the kiln temperature until a small flame was visible through the spy hole. The gas supply would then need to be lowered until the flame just disappeared, and the kiln allowed to continue climbing in an oxidised atmosphere. Forced air is usually supplied by way of a large fan, and kilns fitted with this type of system tend to be a little more automated; when lighting, the system has to deliver the correct fuel to air ratio for ignition to take place. Once the kiln has been lit, however, the same rules apply with regards to balancing fuel and oxygen as for those of naturally aspirated kilns.

The difference in clay body colour in pieces fired in oxidation and reduction. The paler piece at the top of the image is oxidised whereas the lower piece had a slight reduction. Both pieces were fired to 1100°C (2012°F). *Photo: Author*

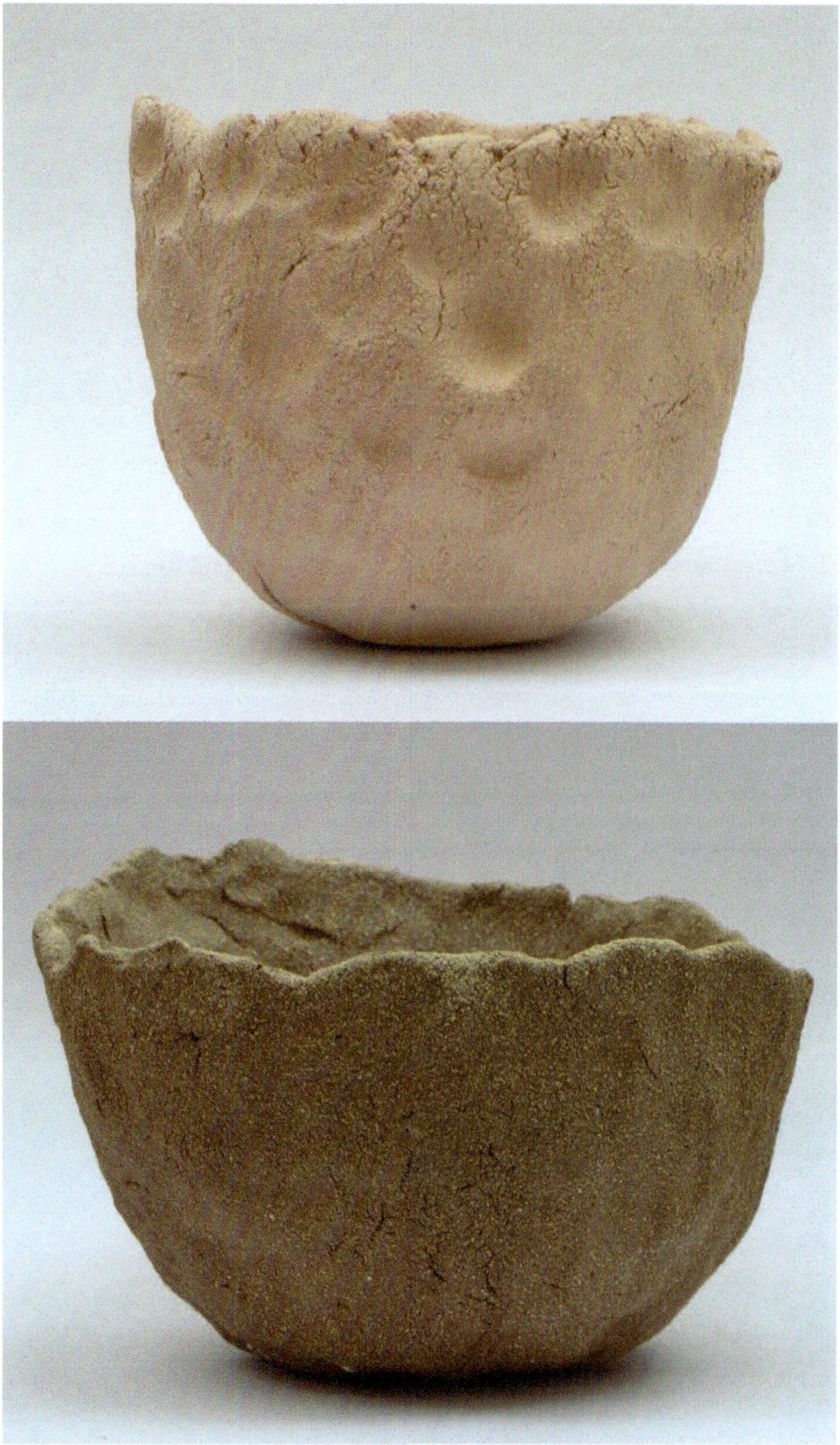

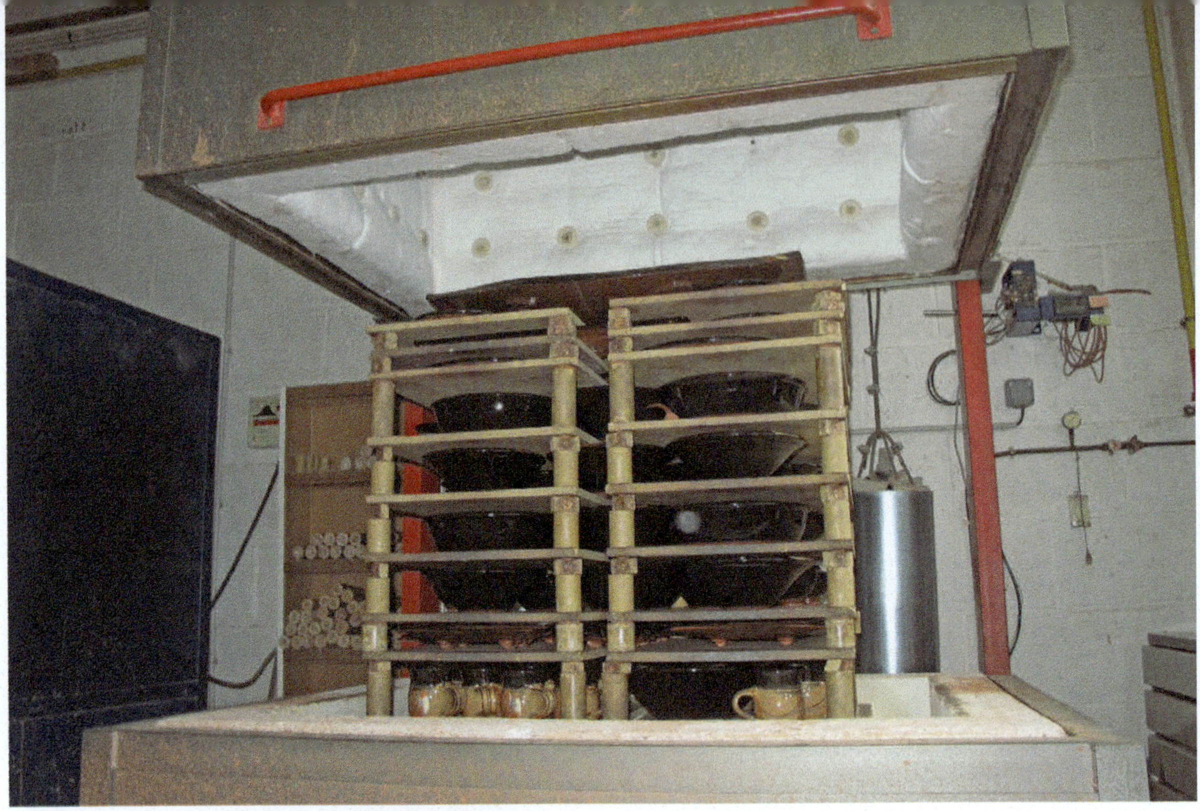

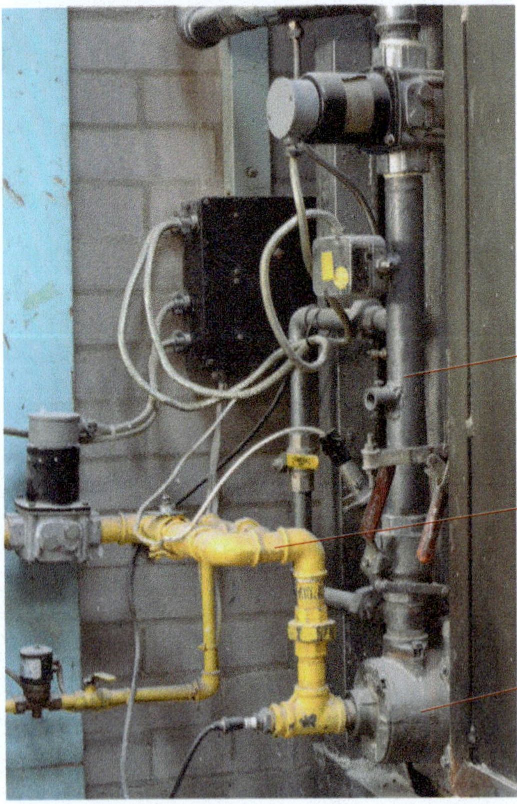

ABOVE: Françoise Dufayard's kiln, ready to unpack. *Photo: Françoise Dufayard*

LEFT: Forced-air, gas-burner system. *Photo: Author*

RIGHT: *Vase*, Françoise Dufayard. Oxidised earthenware. Biscuit firing at 1044C° (1911°F) (Orton Cone 05) for 9 to 10 hours. Glaze firing at 1122°C (2052°F) (Orton Cone 02) for 9 to 10 hours, 66 x 27 cm (26 x 10¾ in). *Photo: Françoise Dufayard*

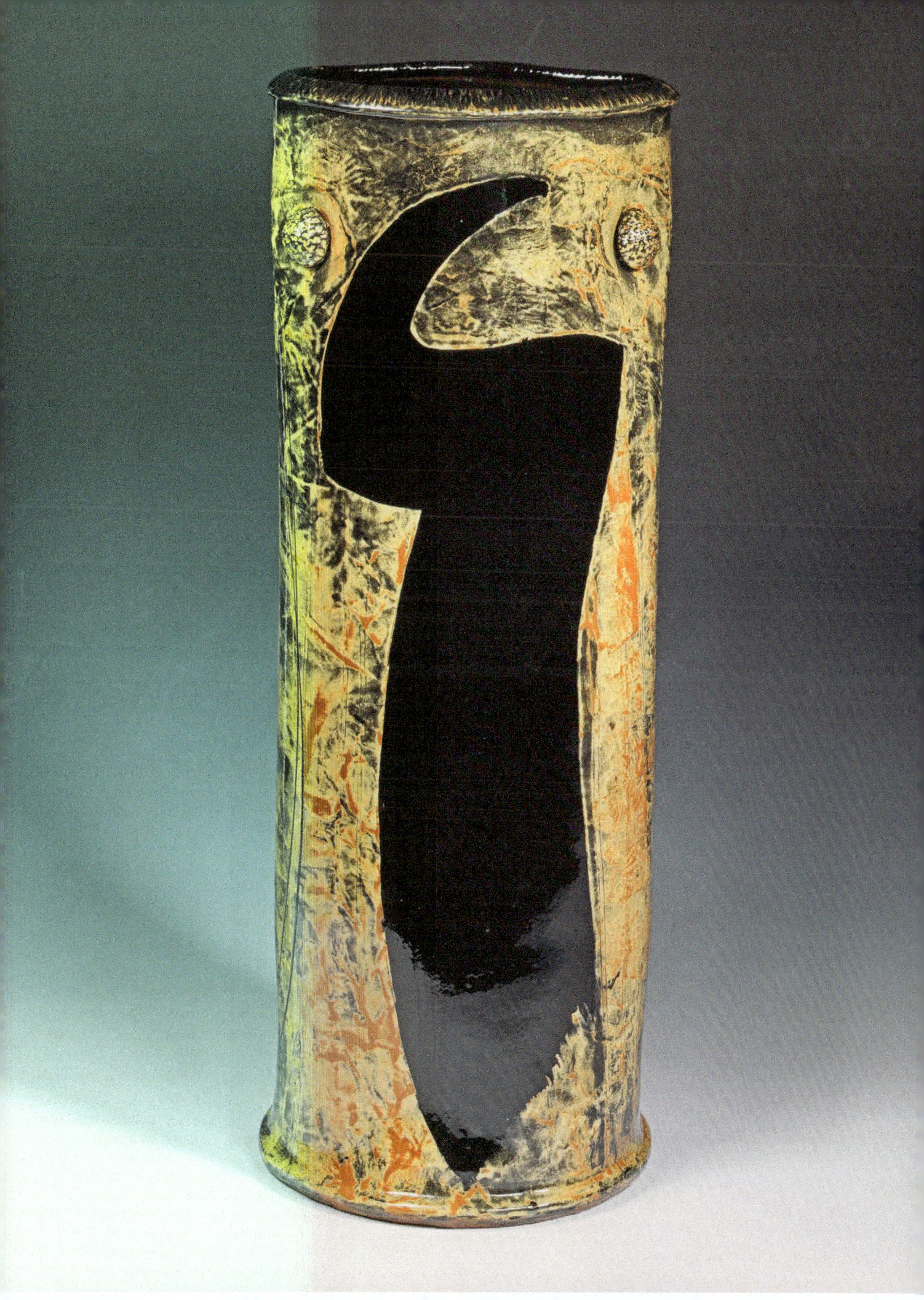

Oil

Oil-fired kilns have similar considerations to those of gas-fired kilns when firing in an oxidised atmosphere. The balance between fuel and air supply is critical, and the available controls are similar in that air intake has to be sufficient for oil to burn efficiently. As with gas kilns the controls usually available are fuel inlet, primary and secondary air inlet and damper position. With most oil-fired kilns the air supply will tend to be delivered by a fan rather than being natural, this allows for good control over combustion. Oil as a fuel tends to be a little dirtier than gas in that it tends to produce more carbon in combustion so a close eye should be kept on the quarls or plates as carbon can accumulate in hard lumps which will ultimately block up these apertures, altering the amount of fuel and air available to be taken into the fireboxes.

The carbon can be removed during the firing by scraping vigorously with a metal pole. A close eye should also be kept on the air inlets to ensure that the accumulation of this carbon does not push the kiln inadvertently into reduction. An added variable with an oil-fired kiln is the type of oil which is being used. As has been mentioned earlier there are several types of oil which can be utilised, for example heating oil, diesel, old engine oil, etc. These may require differing quantities of oxygen in order to combust and may also produce differing calorific output which may need to be accounted for.

Wood-fired kilns

Wood-fired kilns have slightly different considerations to those of continuous-feed, fuel-burning kilns. As wood-kiln firings rely on the periodic addition of fuel through a stoke hole or into fireboxes, they tend to move constantly between oxidation and

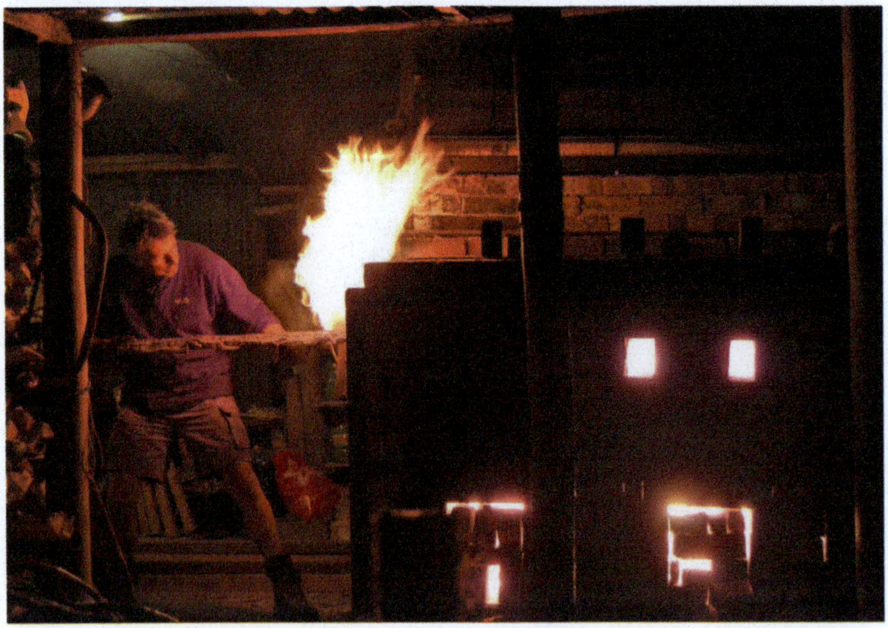

Nigel Lambert stoking his wood-fired kiln. The kiln is a single Bourry box downdraft trolley kiln, constructed from heavy firebricks and lined with HTI (high-temperature insulation) bricks. The term 'Bourry box' describes the way the firebox is configured so that primary air is drawn down through the wood burning on the grate and then into the ware chamber. A trolley kiln is a kiln in which the whole packing space can be pulled out on rails from the kiln structure itself. *Photo: Nigel Lambert*

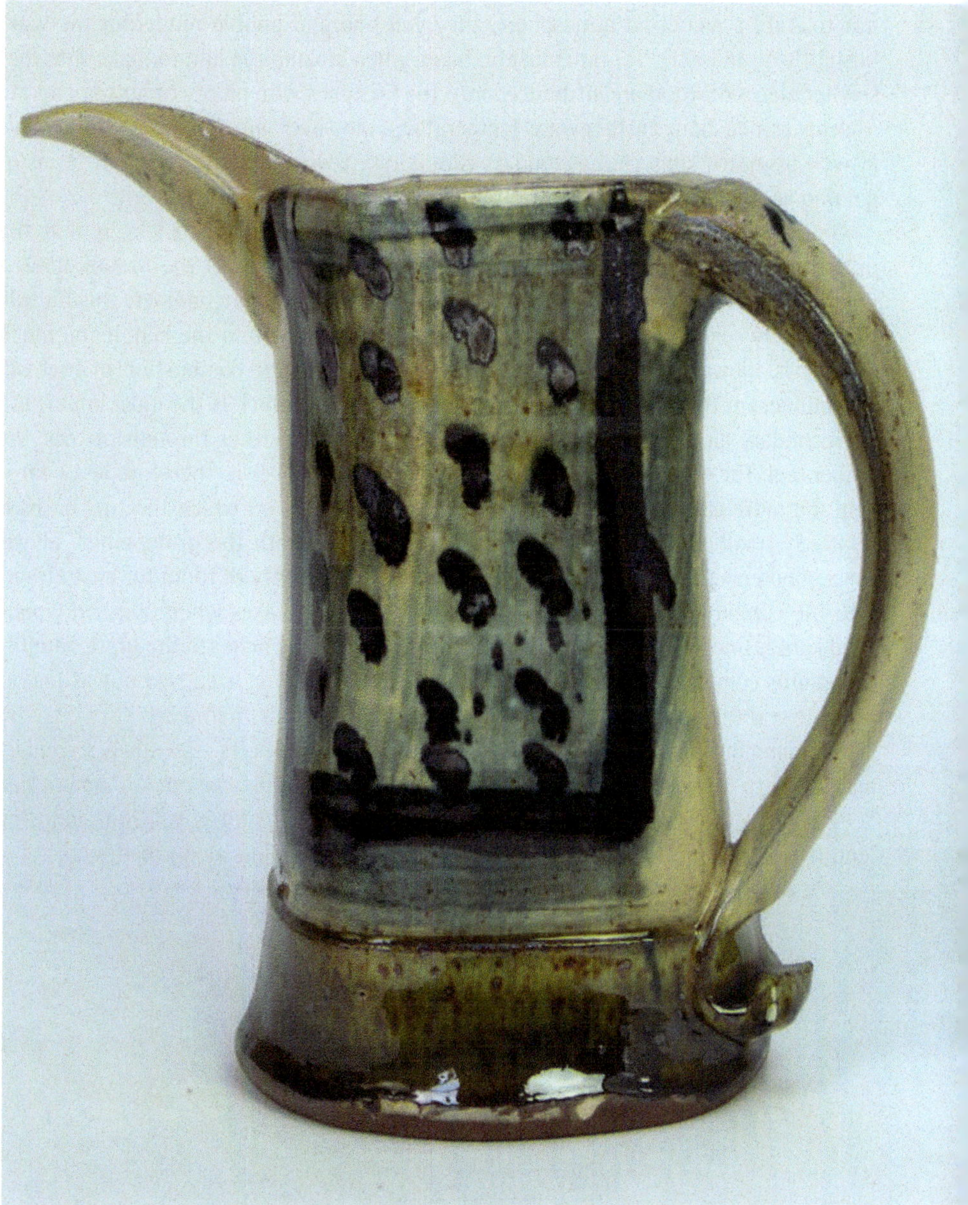

Large Square Jug, Nigel Lambert. Earthenware green and clear glaze with cobalt brushwork, raw glazed and wood-fired to cone 3 flat (1100°C/2012°F) in oxidation, 35 cm (13¾ in). *Photo: Nigel Lambert*

reduction; when fresh fuel is introduced, the kiln will tend to move into a more potentially reducing atmosphere as oxygen is used up in combustion, as the fuel combusts, and the ratio of fuel to oxygen shifts, then a more oxidised atmosphere will be produced. This can be controlled by careful monitoring and adjustment of the primary and secondary air inlets. One tends to find that different wood-firing kiln designs are used depending on the type of firings to be carried out.

Wood-fired kilns designed to fire earthenware will have the capability to introduce more air into the fuel – air mix, this is because lead-based earthenware glazes will not tolerate a reduction atmosphere, they will boil and bubble rendering the work unsightly or unusable. So, care must be taken when stoking the kiln to make sure that smaller pieces of wood are introduced into the fireboxes with plenty of air to generate a clean combustion. Earthenware temperatures, however, are fairly easy to achieve in well-insulated kilns, and as long as care is taken when stoking, firing in terms of gaining and controlling temperature should be reasonably straightforward.

One other consideration with regards to firing in a wood-fired kiln, is that the primary air source is not one which is particularly desirable. As the firebox door is opened to introduce the fuel, cool air is also drawn into the chamber; this air will be relatively cold compared to the ambient temperature within the kiln. If too much cool air is allowed in, any temperature climb caused by the combustion of fuel will be undone. For this reason, it is the secondary air supply which is the most important.

Secondary air is the air which is drawn into the kiln either through or over the ember bed. The ember bed preheats this air as it passes over it, therefore less heat is lost and more gained. So when stoking the kiln, the area on which the fuel burns is relatively small compared to the area of the space beneath the grate, which allows for a good ember bed to build and for there to be still plenty of room for air to travel into the combustion area. One of the most common mistakes when building wood-fired kilns is underestimating the amount of space required beneath the grate, roughly two-thirds compared to one-third above. Stoking, therefore, is carried out as fast as possible and the burn controlled through the undergrate air channels.

It is common for wood-fired kilns to have plenty of 'mouse holes': essentially secondary air inlets into the combustion area which can be used to adjust the rate of combustion. They tend to allow air both through the coal bed and sometimes will run along the length of the firebox allowing oxygen to enter at specific points along the firebox.

RIGHT: *Punchinello,* Thomas Aitken and Kate Hyde. Thrown and handbuilt. *Photo: Thomas Aitken*

BELOW: *Landscape,* Francoise Dufayard. 65 x 65 cm (25½ x 25½ in). *Photo: Francoise Dufayard*

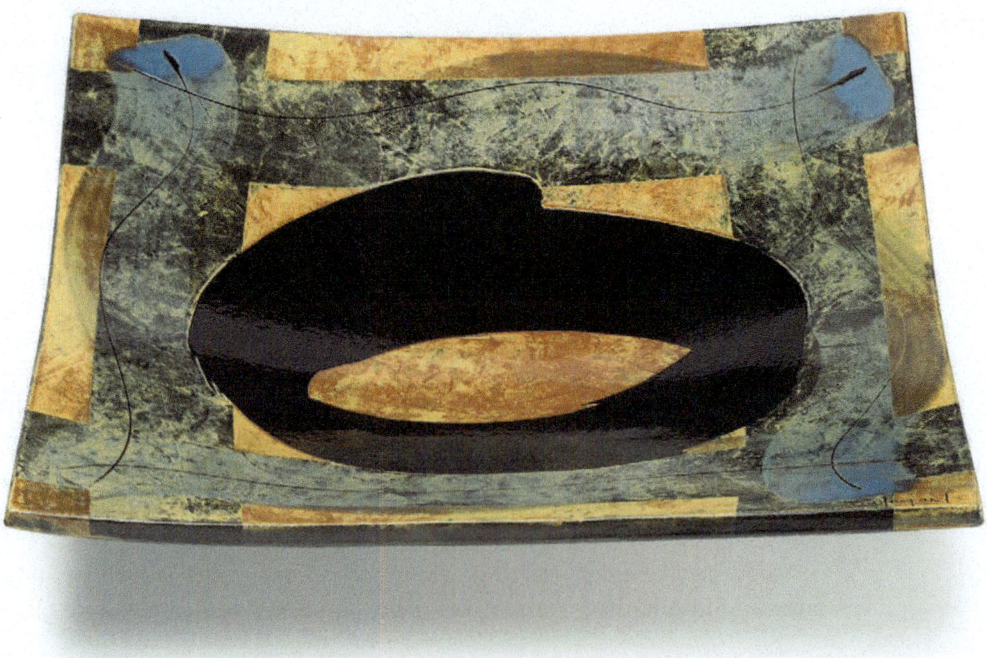

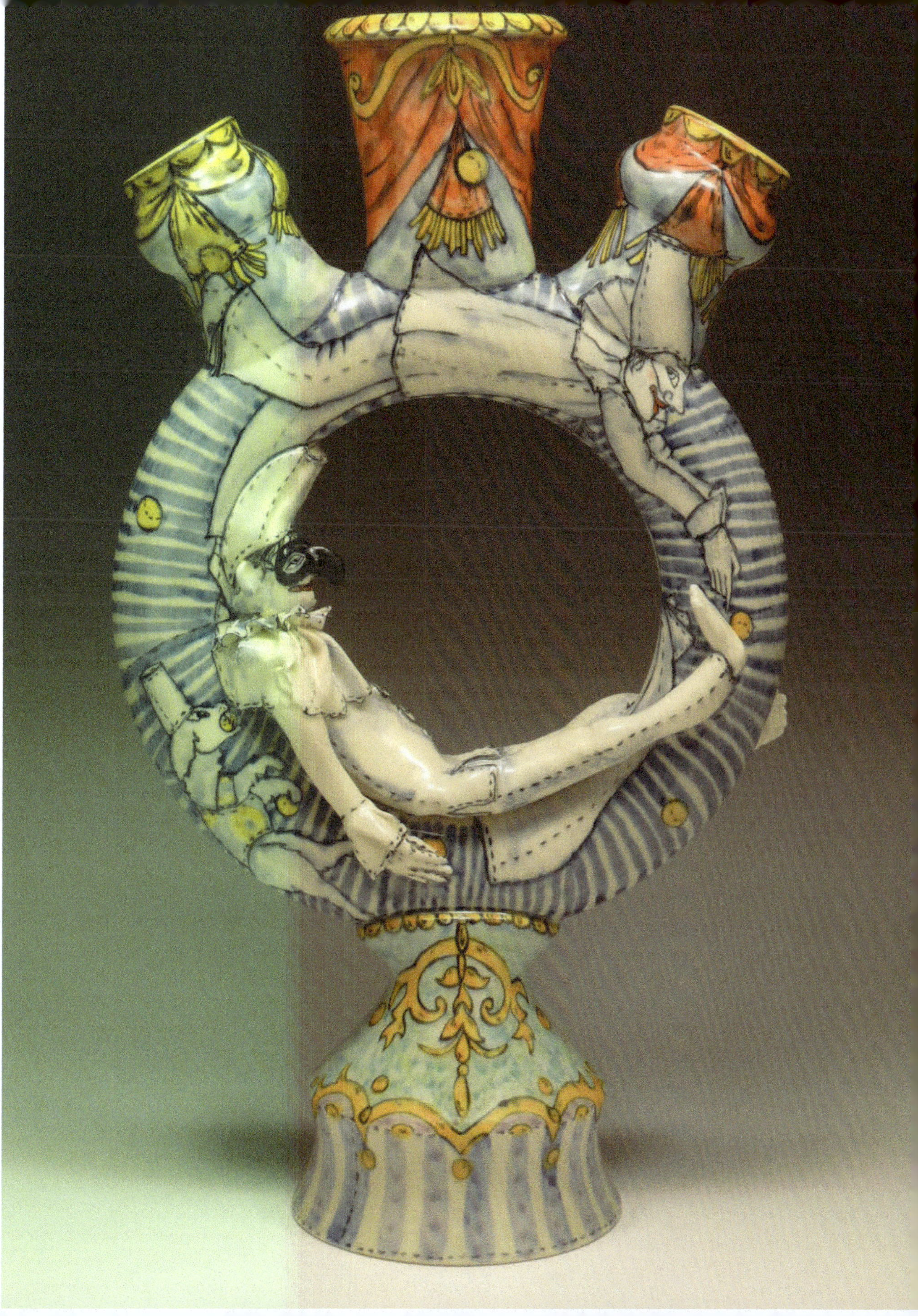

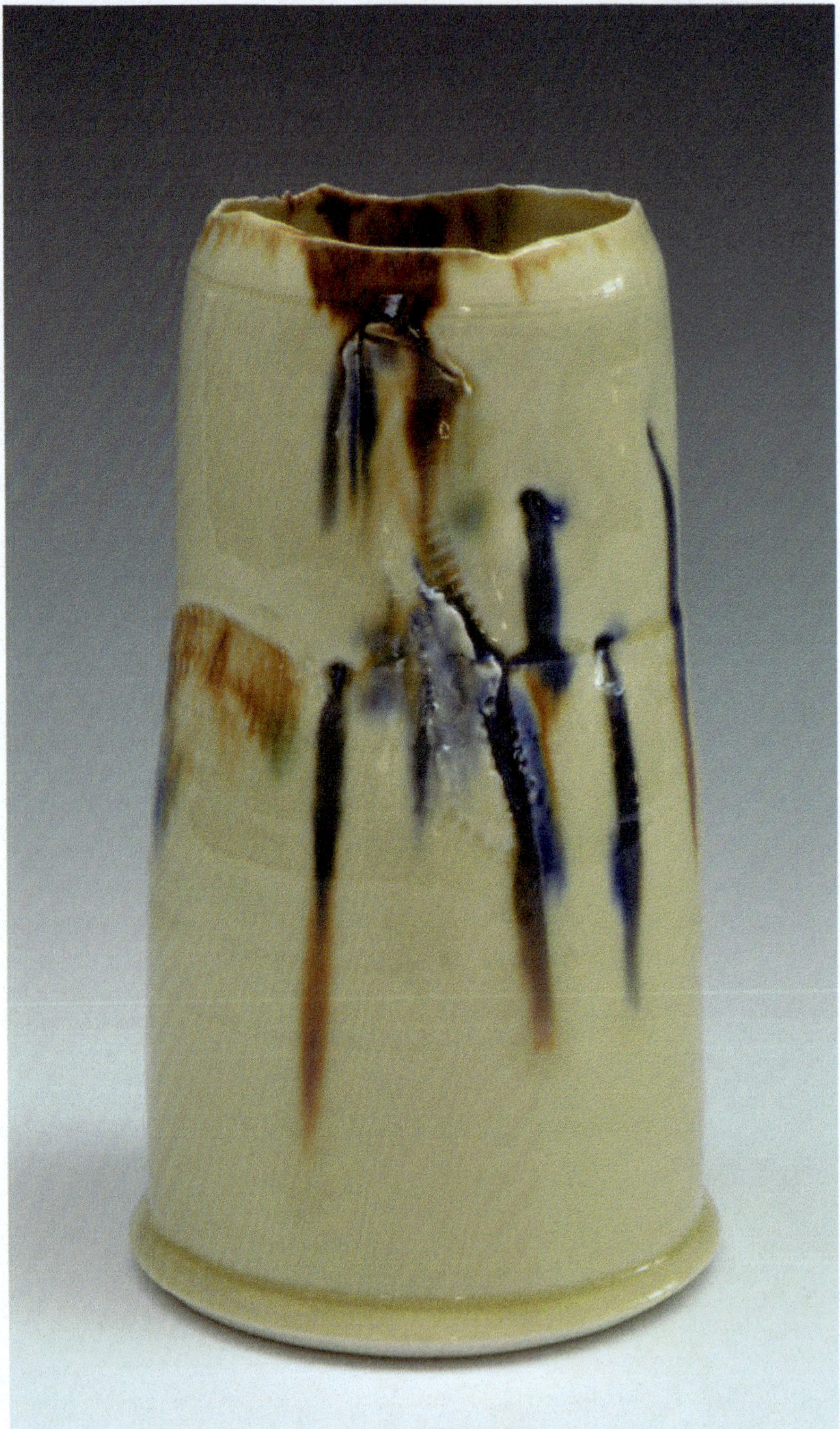

Vase, David Scott. Thrown and altered. Height: 20 cm (8 in). *Photo: Author*

WHY OXIDISE?

One of the benefits of firing in an oxidising atmosphere is that there is no variation; an oxidised atmosphere is efficient heat production. The ceramic materials present in the clays and glazes will have a predictable outcome, which is important if a consistency of production is required. An oxidised atmosphere can produce very crisp colour responses from both clays and glazes, for example what is termed as white stoneware clay will fire white when oxidised, whereas if it were to be fired in a reduction atmosphere, the fired appearance is more likely to be a greyish colour with flecks of iron.

Higher-iron clays when fired in an oxidised firing will give a creamy colour and when fired in a reduction atmosphere the fluxing of iron will create a deep brown. Certain glaze-colouring oxides will also have varying results depending on what atmosphere they are fired in. A small amount of iron when mixed into a stoneware glaze will give a pale honey colouration in oxidation, whereas if fired in a reduction atmosphere, the glaze will be a soft green/blue: a classic celadon glaze. Higher quantities of iron will produce blacks in oxidation and will tend to give black breaking to brick red on edges (tenmoku glaze) in reduction. Similarly with copper, oxidation will give greens and reduction will give pinks and reds.

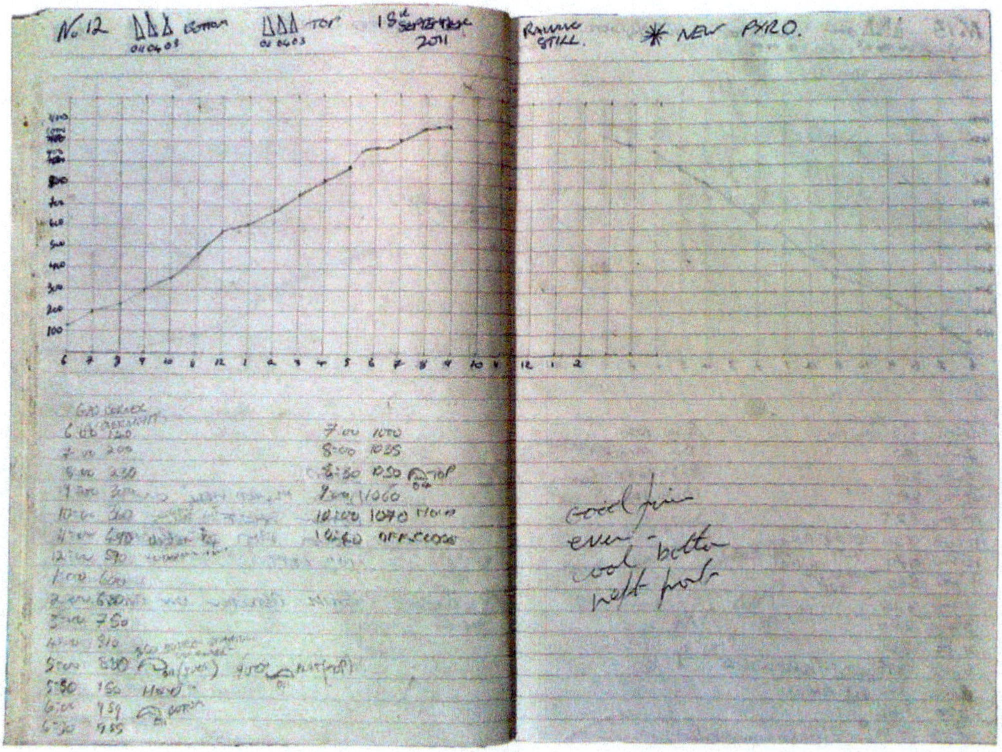

Nigel Lambert's kiln-firing log. Illustrating graph and accompanying notes. This type of recording can be very important when attempting to keep firings consistent especially in kilns which are manually fired. *Photo: Nigel Lambert*

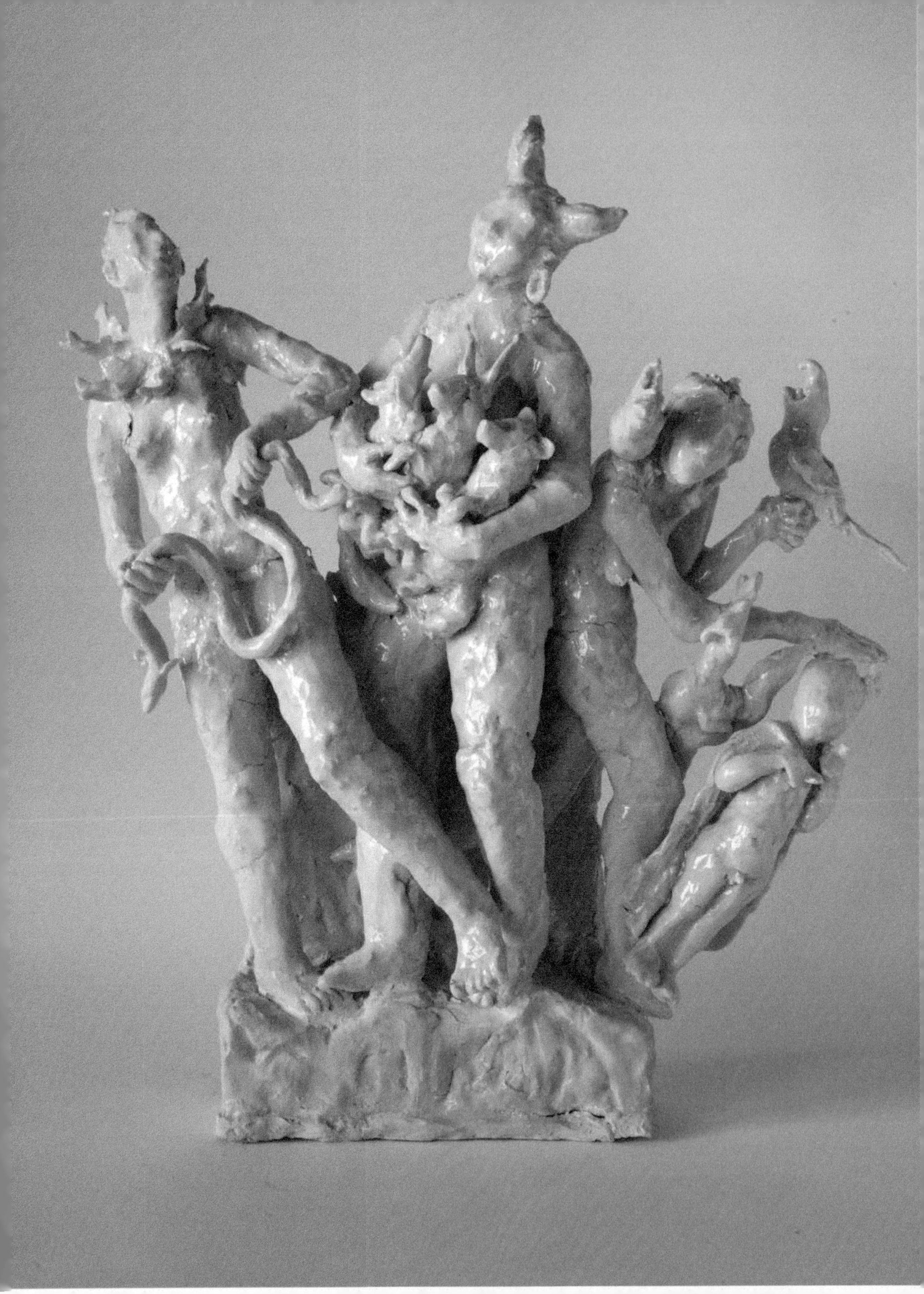

4

Firing in reduction

A reduction atmosphere can be described as one in which the chamber of the kiln either has more fuel available than there is oxygen to burn it efficiently, or that the oxygen supply to burn the fuel is restricted. The action of a reduction atmosphere on the ceramic materials being fired can be very dramatic, adding richness and depth to clays and glazes.

There are generally two periods of reduction within a firing which will give the most visible results. The first occurs around 900°C to 1060°C (1652–1940°F); this is when the kiln is hot enough for reduction of ceramic oxides within the clay body to occur and reduction will generally be started somewhere around this stage of the firing. Iron within the clay is reduced from Fe_2O_2 to FeO, and this occurs due to a build-up of carbon in the kiln due to inefficient combustion. This carbon, in the form of carbon monoxide is very oxygen hungry, and liberates oxygen from ceramic and metal oxides in order to become carbon dioxide. The most abundant metal oxide to be affected by this reduction atmosphere is iron. Iron becomes a vigorous flux when in its reduced state, and can clearly be seen as dark speckles on the surface of the ceramic and will readily bleed through glazes.

It is essential that any kiln which is intended to be used for, or is capable of, firing in reduction is well ventilated. Gases from the flue should be extracted to the outside of the kiln shed and a good supply of fresh air allowed in.

How much the clay is visibly altered by a reducing atmosphere will depend on the amount of iron bound in that clay. A white stoneware body for example will tend to fire grey, possibly with some speckling, as there may be only a small amount of iron in the ball clays used to formulate the clay – the speckling effect is created by the iron contained within any grog in the body. If a very high-iron clay (a red marl to give an extreme example), is reduced at stoneware temperatures, it is quite capable of melting completely to a glaze due to the very large amounts of iron which it contains reducing and acting as a strong flux. Care should be taken that reduction is not started too early. If this happens there is a strong chance that any unburnt organic matter or unliberated gases in the clay will not be able to burn out, due to the fusing of the clay surface, and this will cause the body to bloat.

There are exceptions to this rule. When firing carbon-trapping glazes such as some shinos, starting heavy reduction at around 850°C (1562°F) can be beneficial, allowing larger carbon partilcles to build up on the surface of the glaze which are subsequently bound to the surface of the glaze by the early fluxing of soluable materials.

The second window of reduction happens once the glaze starts to become molten. The same principles are involved, but this time it is the iron in the glaze which is reduced. This can induce some profound colour changes in some glazes. Low-iron

LEFT: *Potato Boy and Three Mothers,* Michael Flynn. Gas reduction-fired porcelain. *Photo: Michael Flynn*

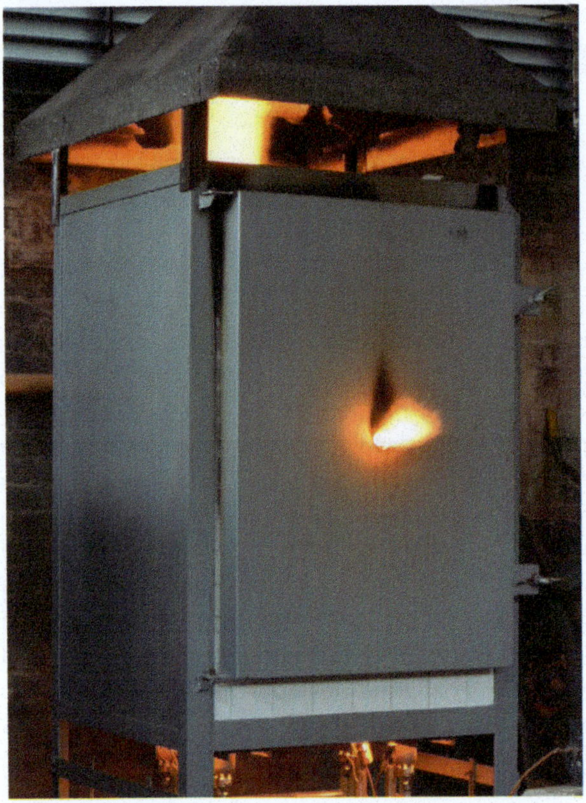

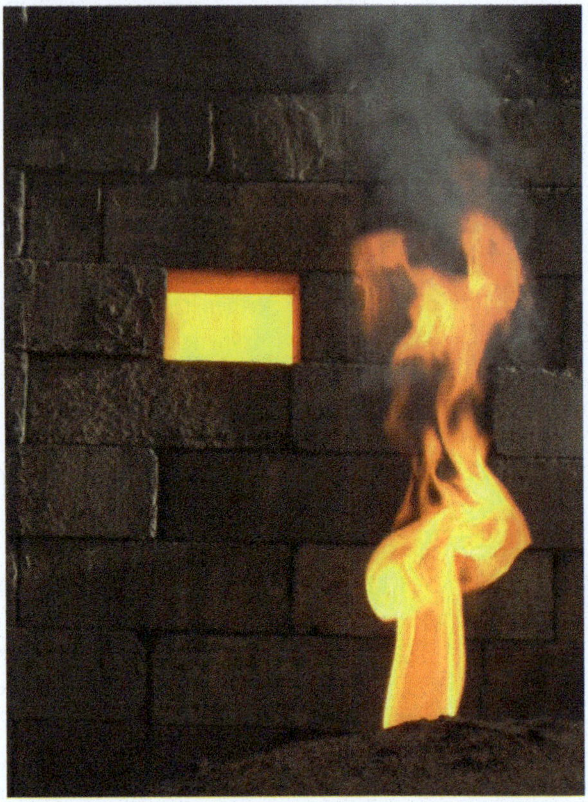

glazes (1–2%), will produce honey-coloured glazes in oxidation and in reduction pale blue/green celadon and chun glazes; 5–6% iron will give rich olive greens; and glazes with 10–12% iron will give classic tenmoku glazes which are black breaking to rust red on edges. Copper is another oxide which can offer strong colour changes: giving greens in oxidation but when combined with tin oxide in reduction providing rich red glazes.

Low-temperature reductions can be carried out with lustre glazes which are applied to pre-fired earthenware tin glazes creating wonderful lustrous surfaces. These are applied as a thin paste over a pre-fired white tin glaze and then fired in a heavy reduction to around 1000°C (1832°F) the lustre is then fused to the white background glaze. Low-temperature raku glazes can also be reduced both in the firing and also in post-firing reduction. These types of glazes tend not to use lead as the primary glaze flux relying more often on alkaline or borax frits. In many cases the degree of reduction will be an individual choice based on an understanding of the kiln and the work being fired, and will be judged by the way the kiln reacts to certain adjustments. In some cases where a specific degree of reduction is consistently required irrespective of the kiln pack or atmospherics which may alter the burner responses, an oxygen probe may be used. This measures the internal combustion atmosphere, so a specific degree of reduction can be repeated firing after firing. They are rather expensive and not commonly used, but are useful if a particular glaze reduction effect is critical.

ABOVE LEFT: 9 cu. ft gas-kiln firing. The flame extending from the spy hole is an indication that the kiln is in a medium to heavy reduction. *Photo: Author*

ABOVE RIGHT: Back pressure from a kiln firing in reduction with flames moving past a passive damper behind. *Photo: Author*

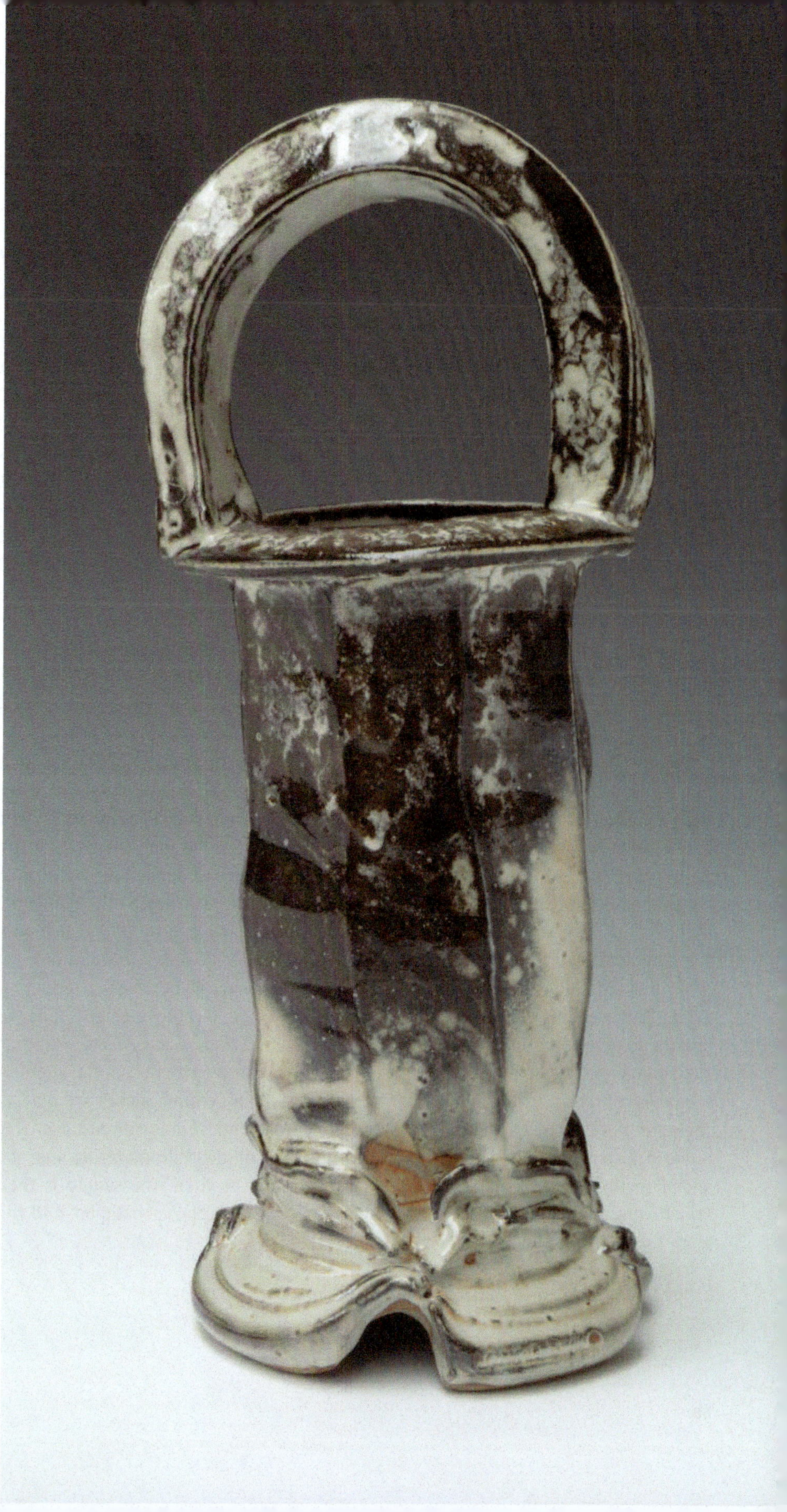

Thrown and altered vessel, Benedict Brierley. This piece has a carbon trapping shino type glaze applied. The black areas are where carbon formed by very heavy early reduction has trapped in the glaze and remained to final glaze maturation, 40 x 15 cm (15¾ x 6 in). *Photo: Author*

REDUCTION IN ELECTRIC KILNS

Having stated in the previous chapter that electric kilns are not used for reduction firings, it is technically possible to carry out glaze reductions in electric kilns. A reduction atmosphere may be produced not just by the over-addition of fuel, it may also be started by the introduction of combustible materials which when ignited by the ambient kiln temperature will produce the carbon monoxide needed to liberate oxygen from the ceramic oxides.

The author carried out glaze-reduction firings in an electric kiln successfully for several years when no fuel-burning kiln was available. The technique is best in stoneware firings and when reduction is started close to the kiln's top temperature. The atmosphere generated tends not to be strong enough to produce a body reduction. The kiln used was a 4 cu. ft top-loader kiln with a spy hole in the lid and two spy holes down the side. When packing the kiln, small trays were positioned on shelves just below the side spy holes and below the top vent. Towards the end of the firing, small slithers of wood were pushed through the side spy holes into the trays, and charcoal was dropped through the top vent into the bowl. As soon as combustibles were added the bungs were replaced; this was maintained for about an hour and produced some striking glaze modifications.

If reduction firings are to be experimented with, the most important consideration is that the kiln is well ventilated, as if it were a fuel-burning kiln, with effective extraction of fumes out of the immediate kiln environment, and plenty of natural ventilation around the kiln.

Electric kiln elements build up an oxide layer over their outer surface which can be seen after a few firings; new elements will be bright and shiny but after several firings will become dull grey in colour – this is the oxide layer on the surface of the element and inadvertently acts as protection for the elements. When the atmosphere shifts into reduction, however, an easy source of oxygen (as well as the ceramic oxides in the work) is also the oxide layer on the surface of the elements. Each firing in reduction will erode this layer, eventually exposing the elements themselves to the corrosive effects of the atmosphere. In order to maintain the layer, it is advisable to carry out several oxidised firings in between a reduction firing. This enables the oxide layer to be replenished on the elements and so maintains the protection.

Saggars may also be used in electric kilns to contain a reduction atmosphere. A saggar is a fireclay or coarse clay container into which ceramic pieces are placed. They are more commonly used to protect work from undesirable kiln atmospheres but can also be used to protect the kiln interior. Combustible materials may be added to the saggar and placed around the work to be fired. The saggar is then sealed with a coil of soft clay covered by a kiln shelf or purpose-made lid, and the work fired.

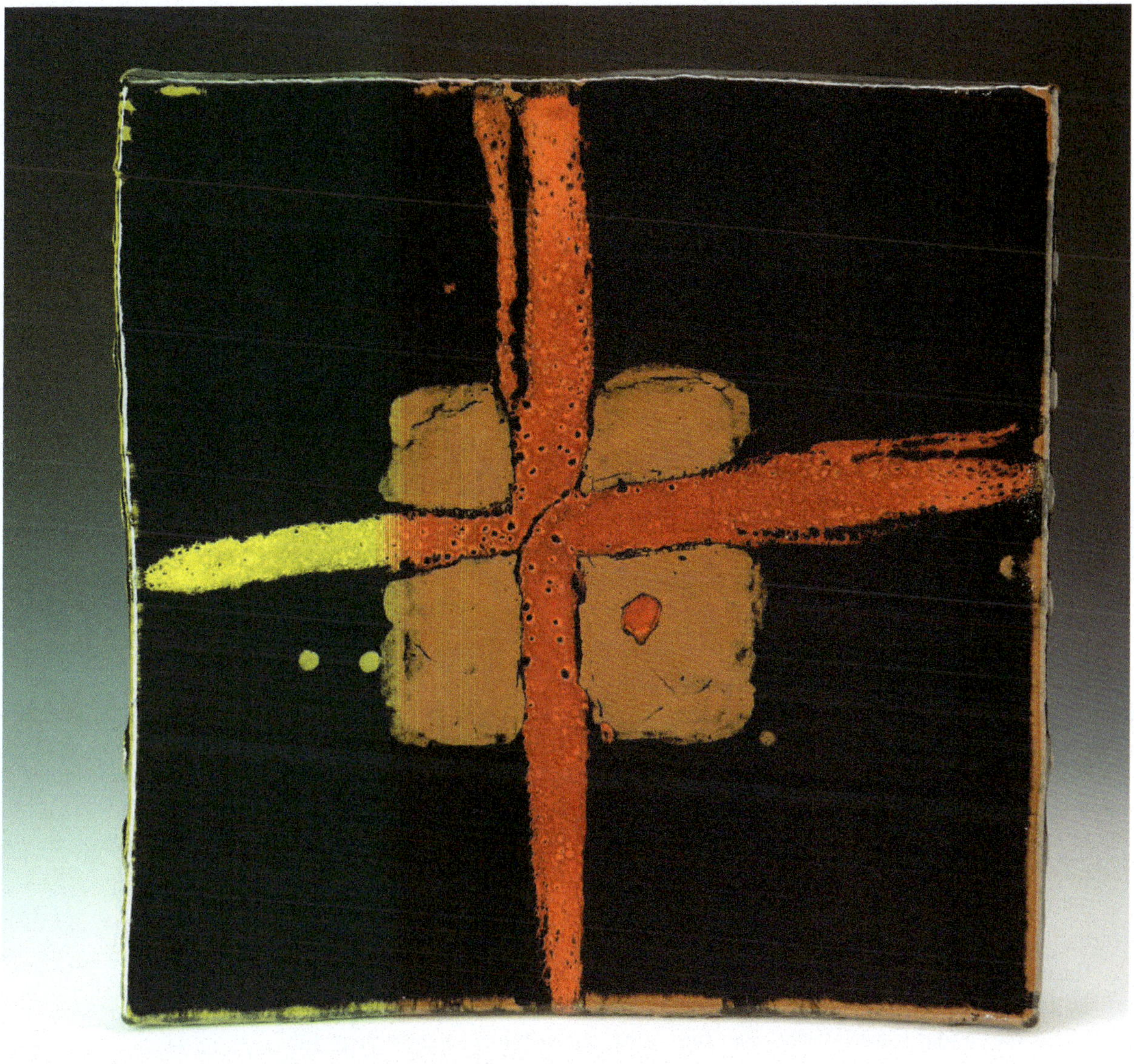

Stoneware platter, James Hake. Fired to 1300°C (2372°F) reduction in a gas kiln, 35 x 35 cm (13¾ x 13¾ in). *Photo: Jonathan Lynch*

REDUCTION IN GAS KILNS

As discussed in the previous chapter, a kiln which is fired on either propane or natural gas is ideally equipped to carry out reduction firings. By the introduction of either an excess of fuel, a restriction of oxygen (primary and secondary air) or a slowing down of the exhaust gases (damper adjustment), a reduction atmosphere may be induced.

Gas kilns used in studio ceramics usually come with either two or four burners. Industrial kilns may have many more running along their length, but in a studio ceramics environment these would be uncommon. Some burner systems will have pilot burners associated with the main burners, some will have a static ignition

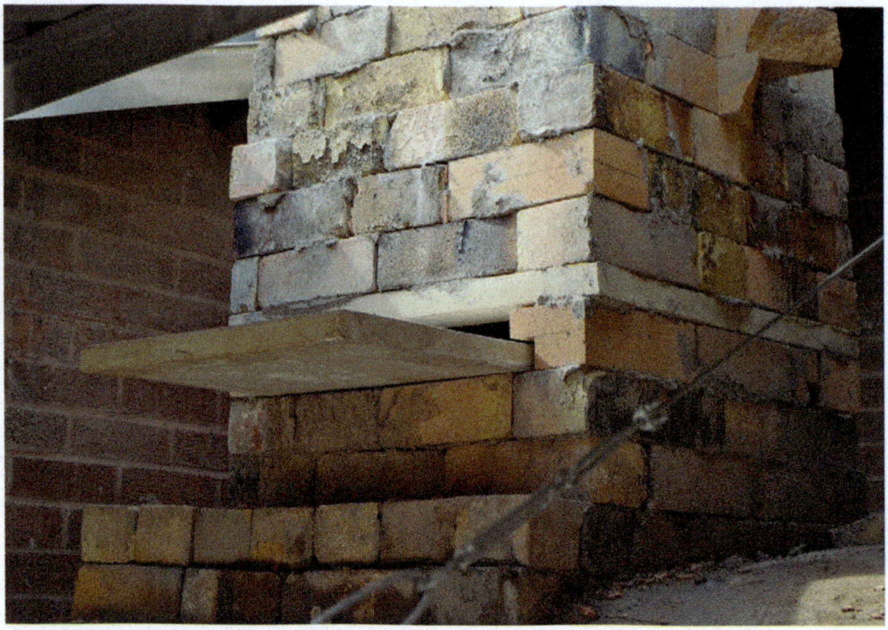

Chimney damper. The damper can be moved in and out to control the degree of reduction inside the kiln chamber. *Photo: Author*

system and others will not. Irrespective of what burner system is associated with a particular kiln the firing principles will be the same. For a reduction firing in a gas kiln, the kiln will be warmed up gently at roughly 100°C/hour (212°F/hour) until the kiln reaches around 980°C (1796°F). At which point one begins to reduce is down to personal preference: for example, when firing carbon-trapping glazes such as high-soda shinos the reduction would be started around 850°C (1562°F), in order to build up fine carbon on the glaze surface before the soda crystals begin to melt at around 900°C to 1000°C (1652–1832°F). Generally the author would start reduction at 960°C (1760°F), when carrying out a reduction firing which did not require a carbon rich atmosphere.

The degree of reduction applied is also down to personal preference. It is sometimes tricky to gauge how much reduction is being applied and this is especially true of gas-fired kilns as gas is a relatively clean fuel. Most kilns will have a spy hole in the door and all will have a chimney or exit flue at the top of the kiln. As the kiln is being reduced, flame will escape from these openings. When firing the gas kiln here at Loughborough University School of the Arts I know that having a 3-inch flame coming out of the spy hole will give me a steady, not too strong reduction. If firing for a heavy reduction I would want to see an 18-inch flame from the exit flue plus an 8-inch flame from the front spy hole.

If too heavy a reduction is applied at these relatively early stages of the firing, the kiln will not climb in temperature because it is not able to burn enough fuel to generate the heat to raise the temperature. Therefore a balance must be gauged between the degree of reduction and the speed of the temperature climb.

RIGHT: *Tall Vessel*, Marcio Mattos. Reduction-fired in his gas kiln to 1260°C (2300°F), 40 x 22 cm (15¾ x 8½ in). *Photo: Marcio Mattos*

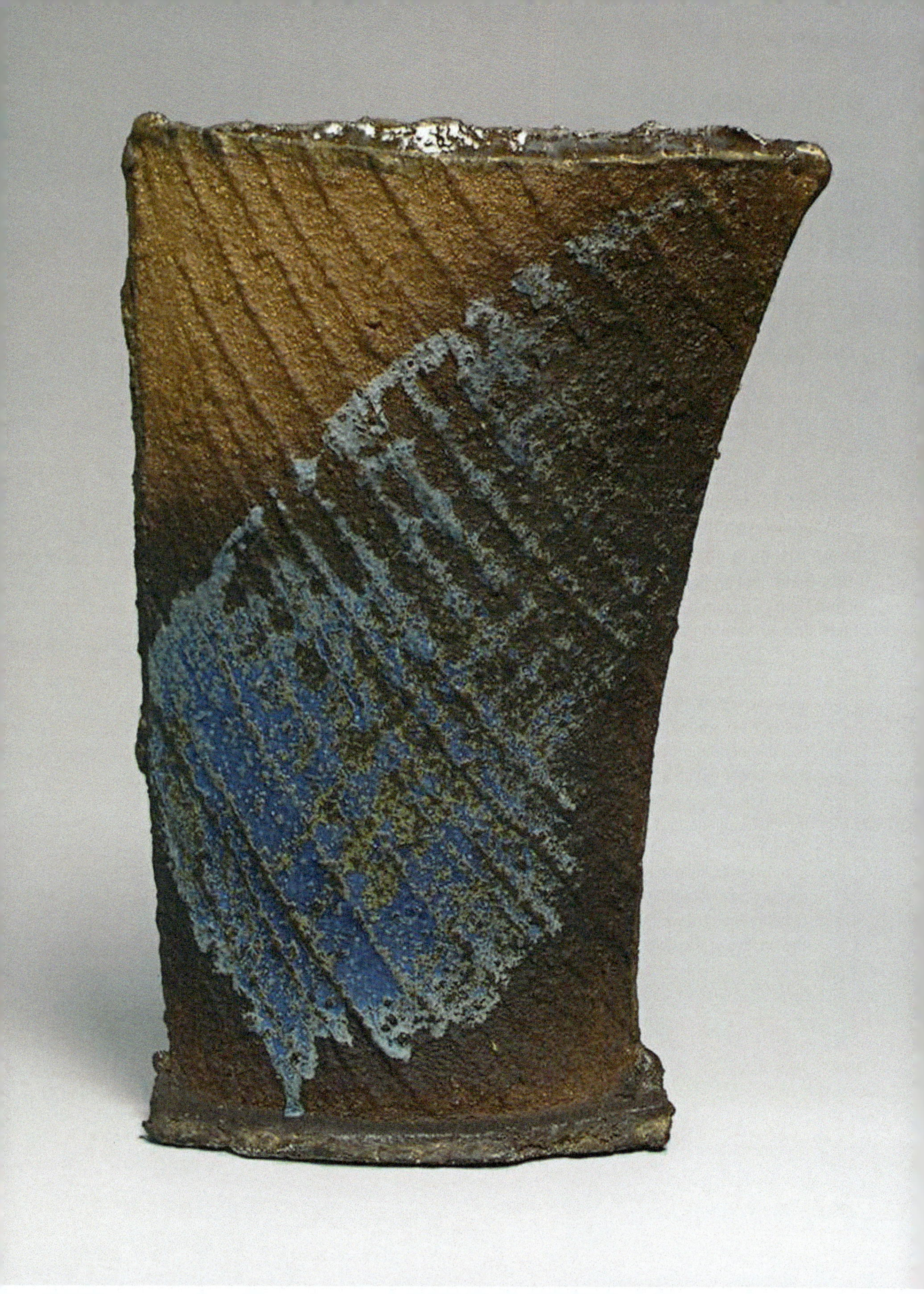

FIRING SCHEDULES

1. 1280°C (2336°F) steady reduction

Cones: 06, 9, 10, 11 (guard)

9am	Light all four pilot burners (door open)
10am	Light two main burners. Opposing: front or back, left or right (Latch door)
12pm	Turn on second pair of burners. All four burners open at 10 o'clock position
2pm	All four burners 11 o'clock position. Cone 06 bending Set chimney damper in 15 cm (6 in): 3-inch flame from front spy hole
4pm	All burners on full Pull damper out to 7 cm (2¾ in), maintain 3-inch flame.
6pm	Cone 9 bending Increase damper 2.5 cm (1 in) in, slightly longer flame from spy
8pm	Cone 10 over. Turn off gas at mains. Close all burners

Finish

2. 1280°C (2336°F) reduction firing with heavy-body reduction, and reduction cooling

Cones: 012, 06, 9, 10, 11 (guard)

9am	Light pilot burners (door open)
10am	Light two main burners. Opposing: front/back, left/right (Latch door)
12pm	Turn on second pair of burners. All four burners open 10 o'clock position
2pm	All four burners 11 o'clock position. Cone 012 bending Start reduction – all burners on full and set chimney damper 15 cm (6 in) in: 8-inch flame from front spy hole. 18-inch flame from chimney. Slight smoke from around front spy bung when in
3pm	Pull damper out 12 cm (4¾ in), 3-inch flame from spy
4pm	Pull damper out to a little to 7 cm (2¾ in), maintain 3-inch flame
6pm	Cone 9 bending Increase damper back to 15 cm (6 in), 7-inch flame from spy, 12-inch from chimney
8pm	Cone 10 over. Push in damper to 20 cm (7¾ in). Large flames from chimney and spy hole. Kiln unable to climb as it is not burning enough fuel so kiln temperature drops with very heavy reduction. This firing will pull out iron to the surface of the glaze and form micro and/or macro crystals in some glazes.

Finish

RIGHT: *Kristoff Stories* (from the series *Trans Forms Plus*, Monika Patuszynska, 2012. Reduction-fired porcelain, height (each piece): 24 cm (9½ in). Fired to 1380°C (2516°F) in a gas-fired tunnel kiln at the tableware factory, Kristoff, Walbrzych, Poland. *Photo: Grzegorz Stadnik*

Stoneware platter, James Hake. Fired to 1300°C (2372°F) reduction in a gas kiln, 35 x 35 cm (17¾ x 17¾ in). James uses a gas-fired kiln constructed from HTI brick, in a steel frame. His firing schedule is: roughly 6 hours to 960°C (1760°F) where reduction starts, 4 to 5 hours of reduction to 1300°C (2372°F), then 30-minute soak. Finally crash cooled to 1000°C (1832°F) and then left to cool naturally. *Photo: Jonathan Lynch*

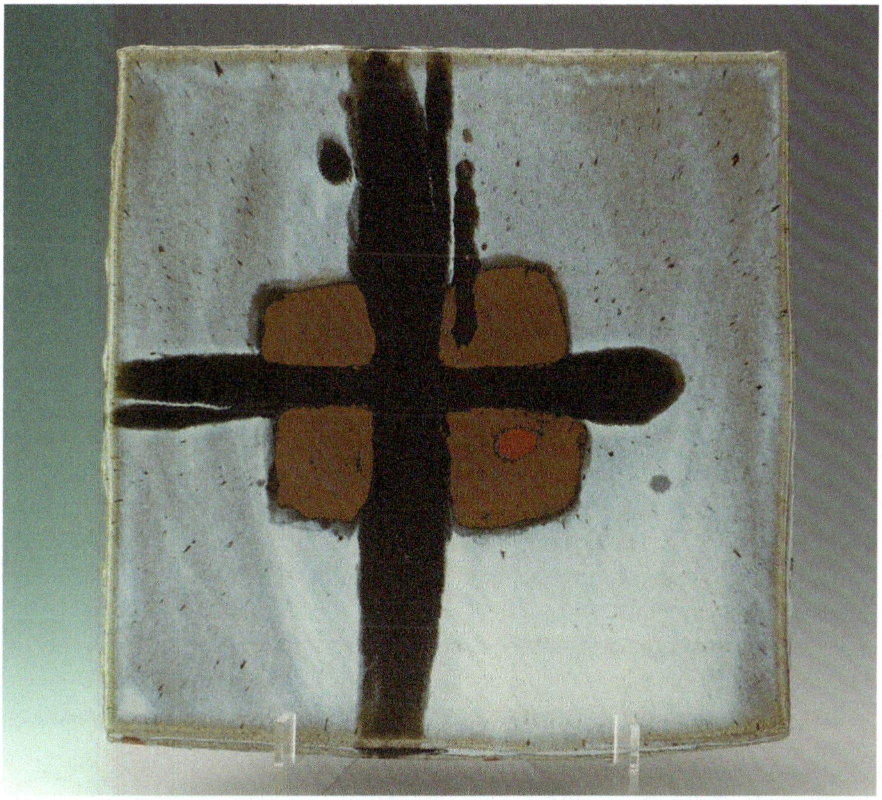

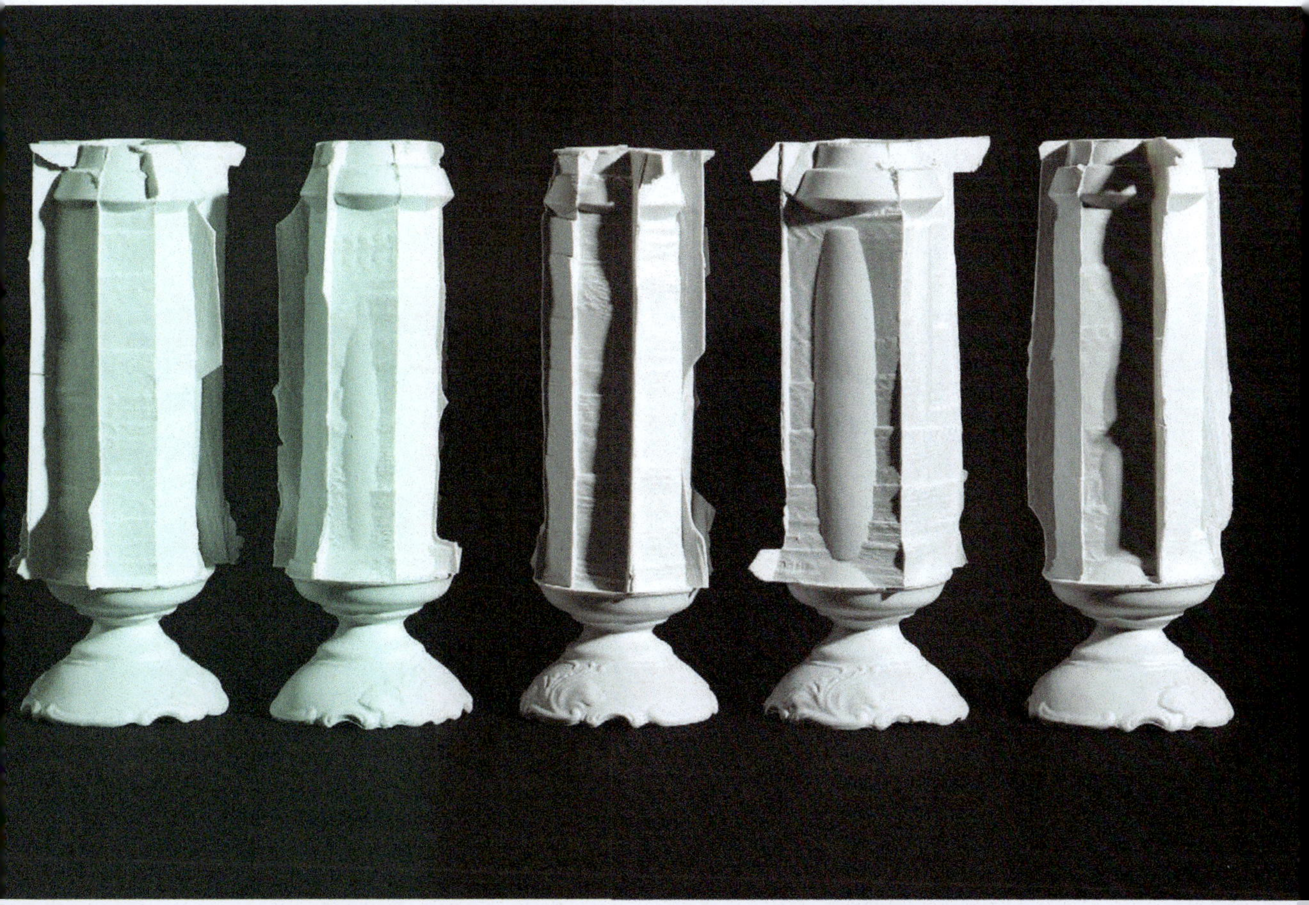

REDUCTION FIRING IN OIL-FIRED KILNS

As has been mentioned in the previous chapter, oil-fired kilns are capable of producing very strong reduction atmospheres. Due to the high levels of soot produced by the inefficient combustion of oil, kilns fired with this fuel are exceedingly good when firing clays and glazes which respond well to 'dirty' atmospheres such as copper-red glazes, chun, celedons and tenmoku. However, on an environmental level this is not so desirable. How the reduction is achieved and the degree of reduction produced relies on similar principles to those applied when firing a gas-fuelled kiln. The controls usually available are primary and secondary air feeds, fuel, and the relative position of the damper.

Oil-burner systems vary slightly, the two main types being atomiser and quarl or drip tray. With the atomiser and quarl arrangement the primary air supply is provided by a fan attached to the atomiser. Burners will have an air valve and an oil valve; by adjusting the relative position of both, a reduction atmosphere may be generated and any secondary air will, depending on the kiln design, be drawn in around the main burner block or through air ports along the side of the firebox. The damper in the chimney can also be used to determine the degree of reduction and is sometimes a more convenient method. If the balance between air and oil are set so that the kiln will climb gently in temperature then the damper can be used to start reduction by altering the fuel–air mix; if the reduction is needed to be reduced then the damper can simply be pulled out a little without the need to rebalance the fuel–air mixture.

Teapot, John Claver. Reduction-fired in his 40 cu. ft oil-fired kiln. John's kiln is a heavy firebrick, sprung-arch downdraft, lined with HTI bricks. Fired using diagonally opposing Dyne oil burners fired with heating oil. The kiln is fired in oxidation to 700°C (1292°F) and then in reduction up to 1320°C (2408°F), shut down and allowed to cool, 20 cm (7¾ in). *Photo: John Calver*

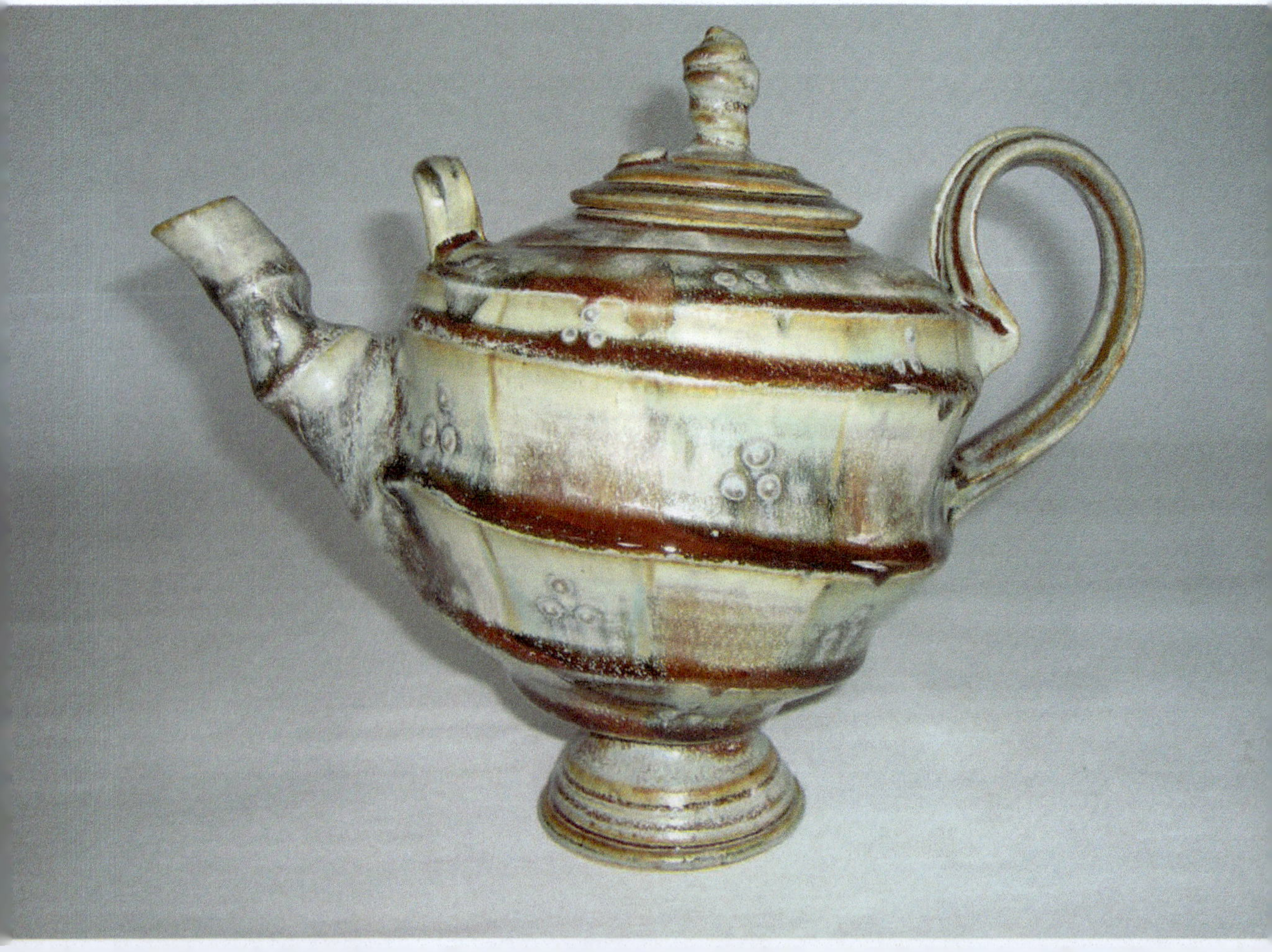

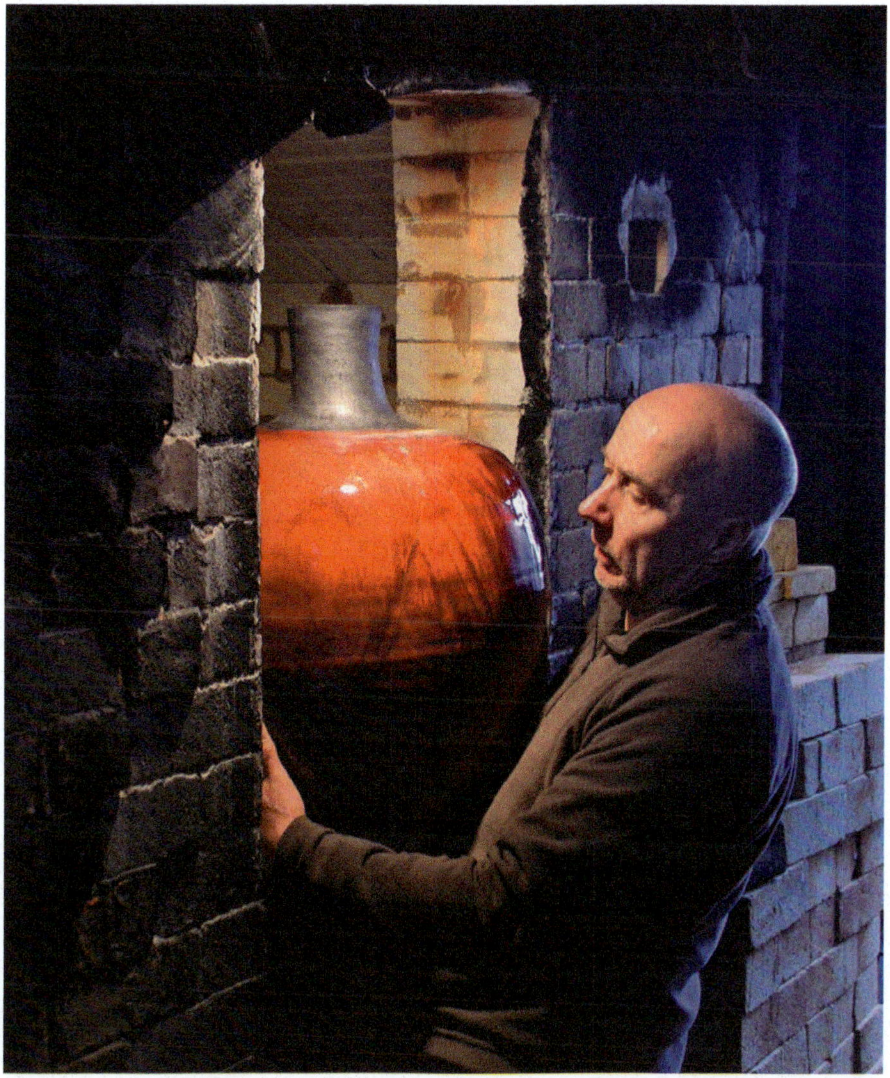

Eddie Curtis unpacking a large copper-red vessel from his oil-fired kiln. *Photo: Eddie Curtis*

With an oil-burner system incorporating drip trays, there are two possible ways of introducing primary air: first, it may be drawn in to the chamber by the use of the draft generated by a tall chimney; and second, a fan may be used as with the venturi burner system. In order to start a reduction either more fuel may be allowed to drip onto the trays, forced air from a fan may be increased or the reduction can be controlled just through the use of the damper and control of the secondary air. In the same manner as with a gas kiln the damper will be pushed in to begin the reduction and the degree of reduction will be determined by how closed off the chimney becomes. In order to lessen the degree of reduction the damper will be pulled out. Again, as with gas kilns it is possible to over reduce the kiln so that it cannot burn enough fuel to climb in temperature, so a balance must be found between the fuel, air and damper position.

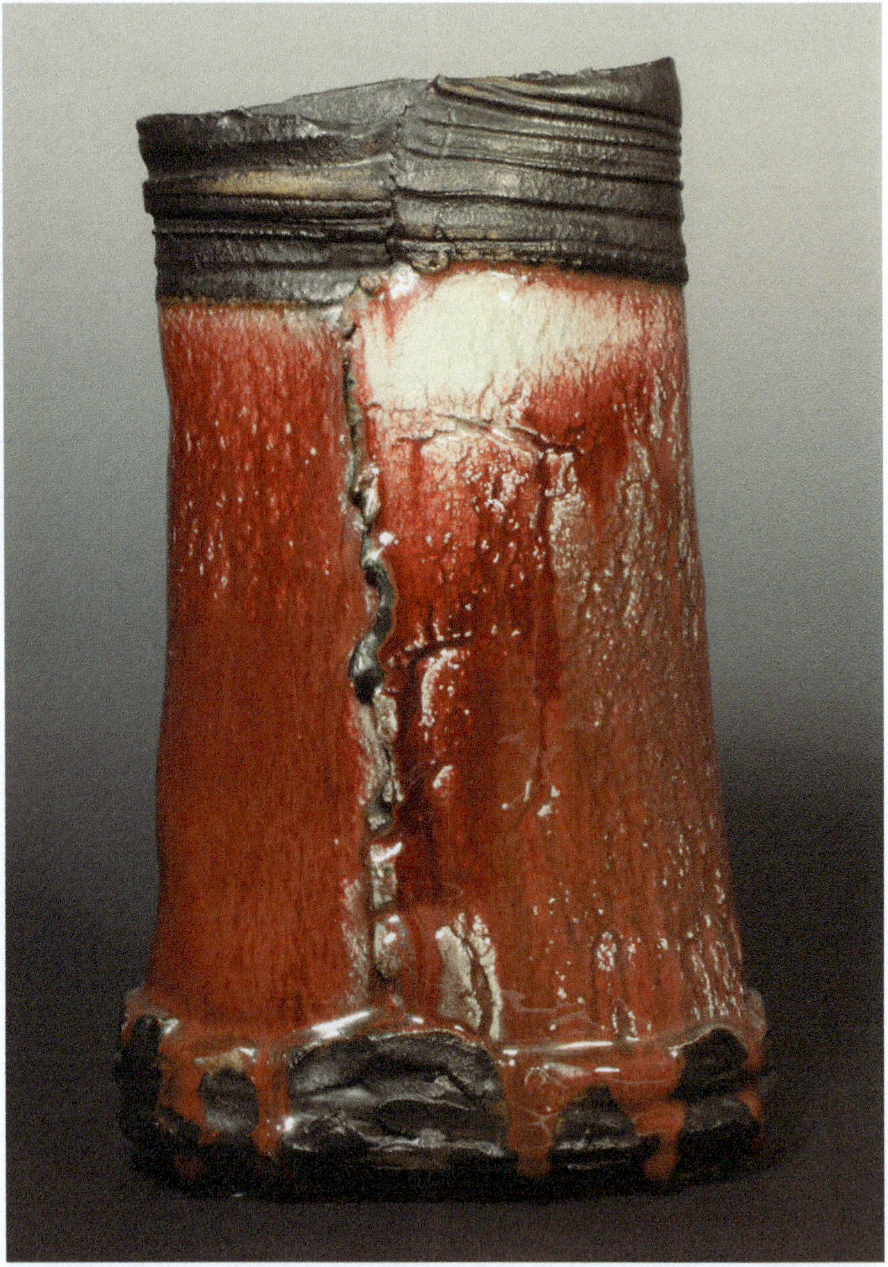

Wrap vessel, Eddie Curtis. Eddie says: 'I work with a large brick-built kiln, constructed by myself over 25 years ago. It stands some seven feet tall and six feet wide and is powered by two oil burners, driven by compressed air. It's noisy, intense and very exciting, a force cajoled rather than controlled, and never tamed. The burners roar, the flames leap out of each tiny crack in the walls of the kiln, and the heat is searing. A typical firing will last perhaps 14 hours and achieve a temperature in excess of 1300°C (2372°F). I tend the kiln for the duration of the firing, making adjustments to the burn rate and controlling the internal atmosphere.'
Photo: Eddie Curtis

REDUCTION FIRING IN WOOD KILNS

A wood kiln will, by the nature of the way fuel is introduced, tend to fluctuate between oxidation and reduction atmospheres. When fresh fuel is introduced, air will be drawn through the ember bed; initially there will be more fuel than there is air to combust efficiently as the fuel burns down, however this balance will change as the ratio of fuel to air changes. In my wood-fired anagama I have a spy hole positioned roughly halfway up the chimney which is viewable from the front of the kiln and when the kiln is stoked above 1000°C (1832°F) a flame is visible passing this hole. This can then be used as a stoking guide to ensure that the chamber is kept in reduction; when the flame drops out of sight I know it is time to stoke the kiln again. By controlling the primary air intakes beneath the firebox grate, the amount of oxygen available for combustion may be controlled. Many wood-fired kilns will also have a damper in the chimney; as the chimney is a fundamental element in the draw of the kiln, the use of a damper in order to induce a reduction can be very effective.

Some kilns will also have passive dampers, which are essentially holes in the chimney that can be opened and closed. When the passive damper is opened, air is drawn into the base of the chimney; this will reduce the pull of air from the front of the kiln which will help to unbalance the air-fuel ratio and promote a reducing atmosphere. As with any other method of reducing fuel-burning kilns, the degree of reduction and at what stages the kiln is reduced will be down to personal preference. Having said this, when I fire I tend to try and keep the kiln in some degree of reduction from about 900°C (1652°F). Because I am using very little applied glaze and am relying on the composition of the clay body to generate colour this reduction cycle is very important.

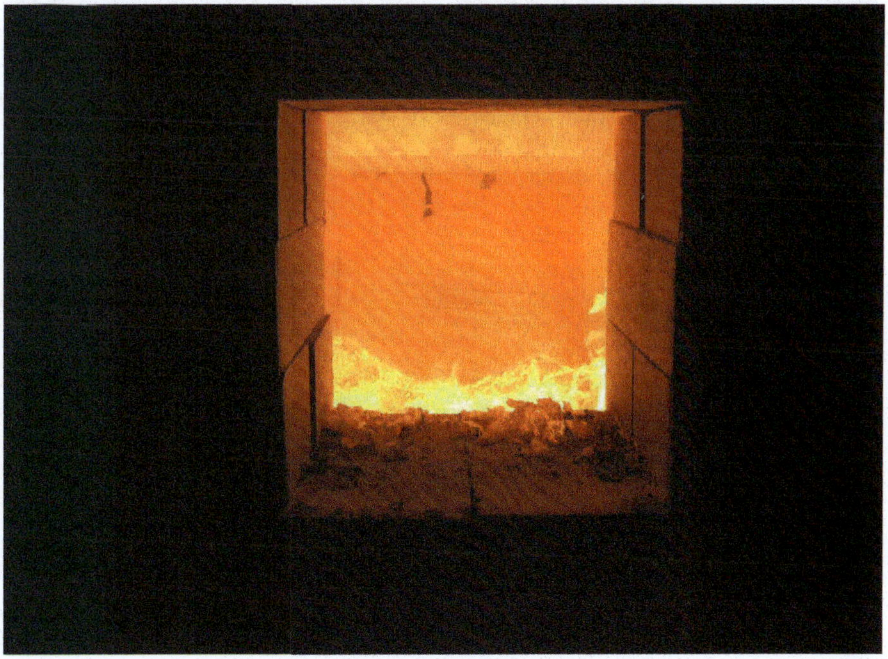

A view through the upper stoking port of an anagama kiln. *Photo: Author*

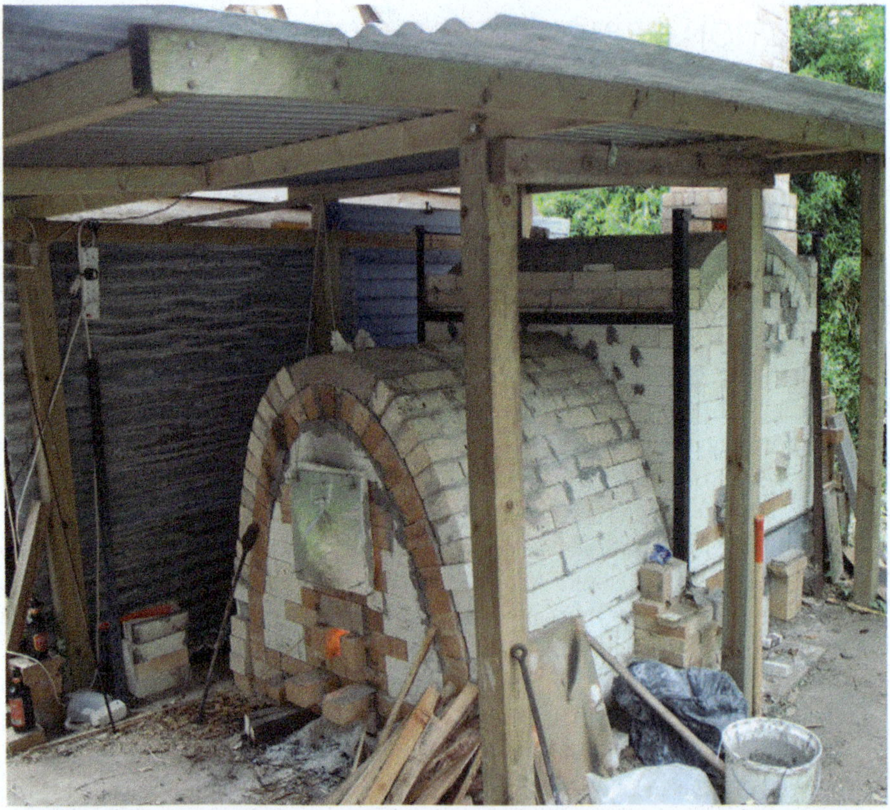

LEFT: Matthew Blakely's crossdraft wood-fired kiln. This kiln based on Dick Lehman's kiln combines a catenary arch firebox/ stacking space with a sprung arch stacking chamber. *Photo: Matthew Blakely*

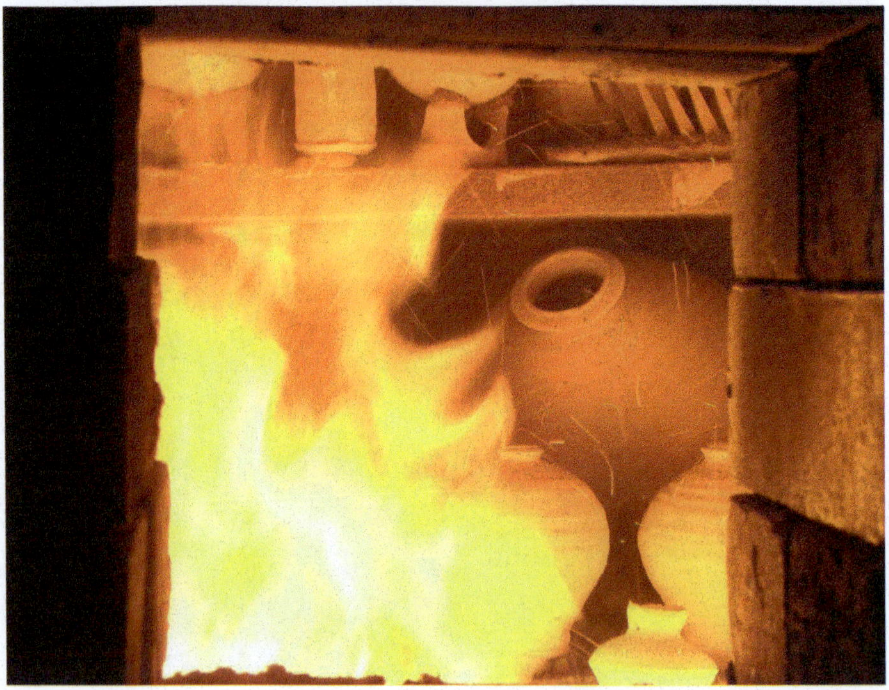

LEFT: View into a wood-fired anagama kiln at the start of a 4-day firing. *Photo: Author*

RIGHT: *Pair of Bottles*, Stephen Parry. Fired to Orton cone 11 (1300°C/2372°F), in his Phoenix-style crossdraft wood-fired kiln, 60 x 25 cm (23½ x 9¾ in). *Photo: Stephen Parry*

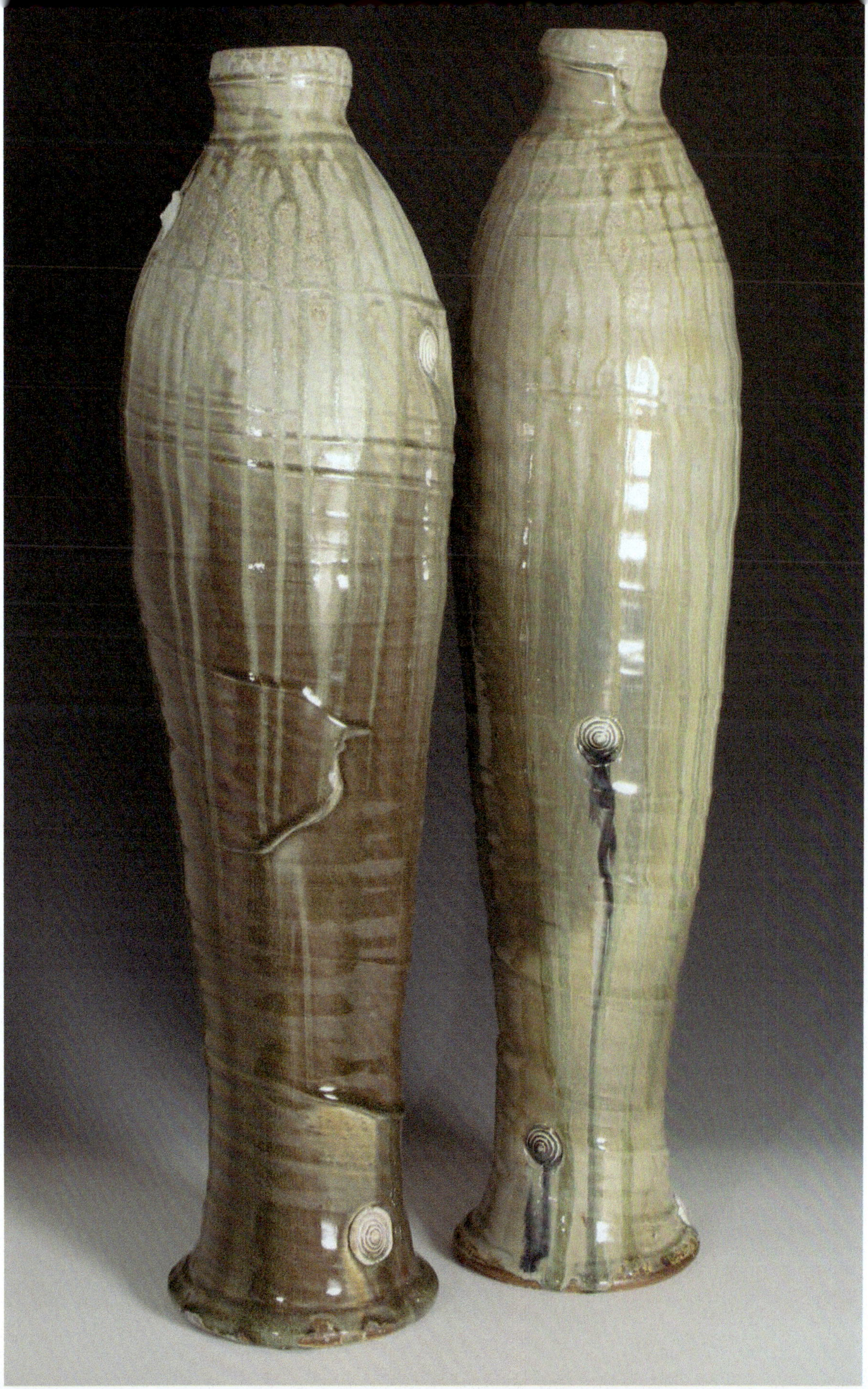

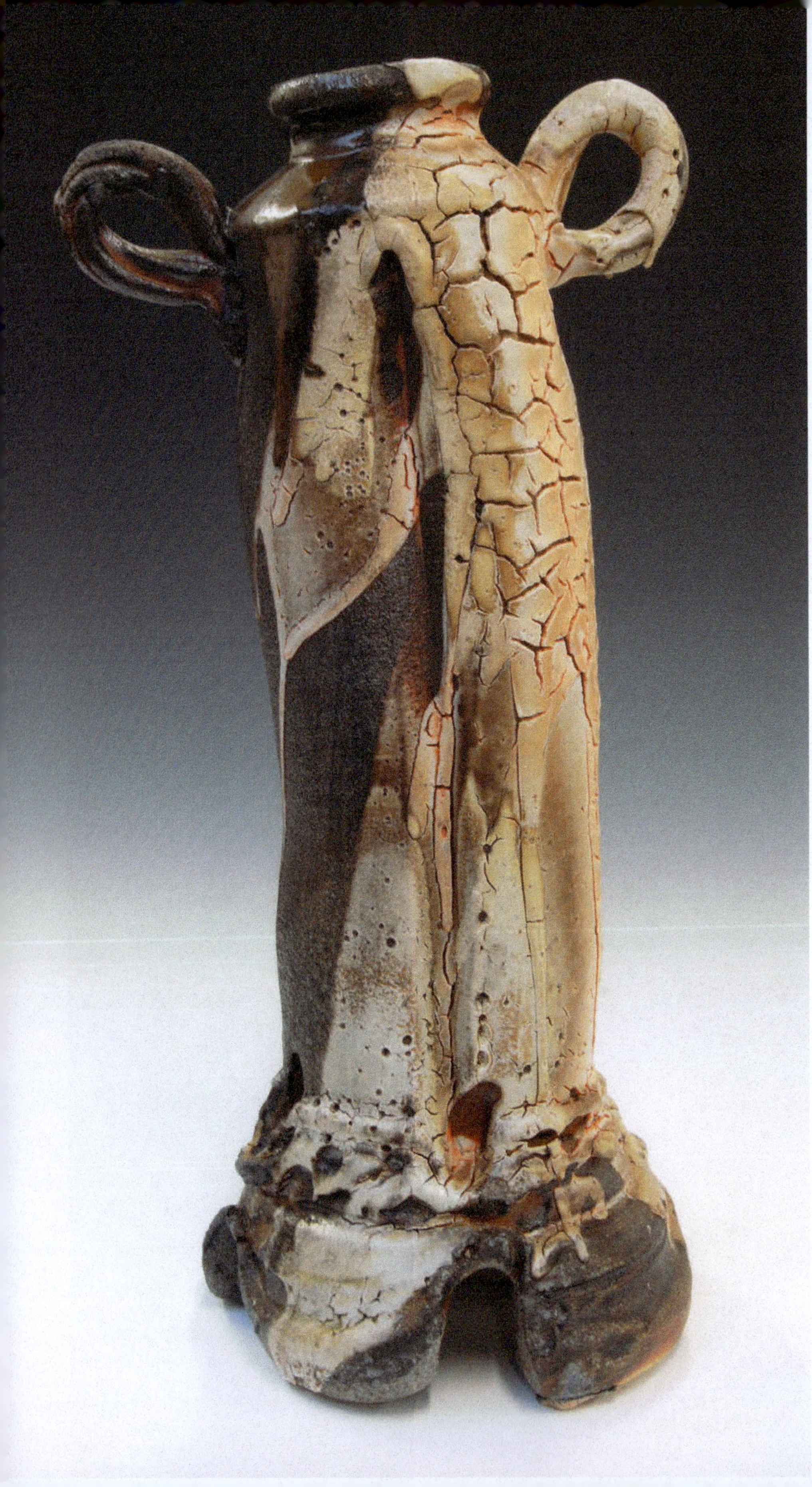

LEFT: *Tall vase form,* Benedict Brierley. Thrown and assembled. Wood fired for 4 days in a crossdraft kiln. Height: 34 cm (13¼ in). *Photo: Author*

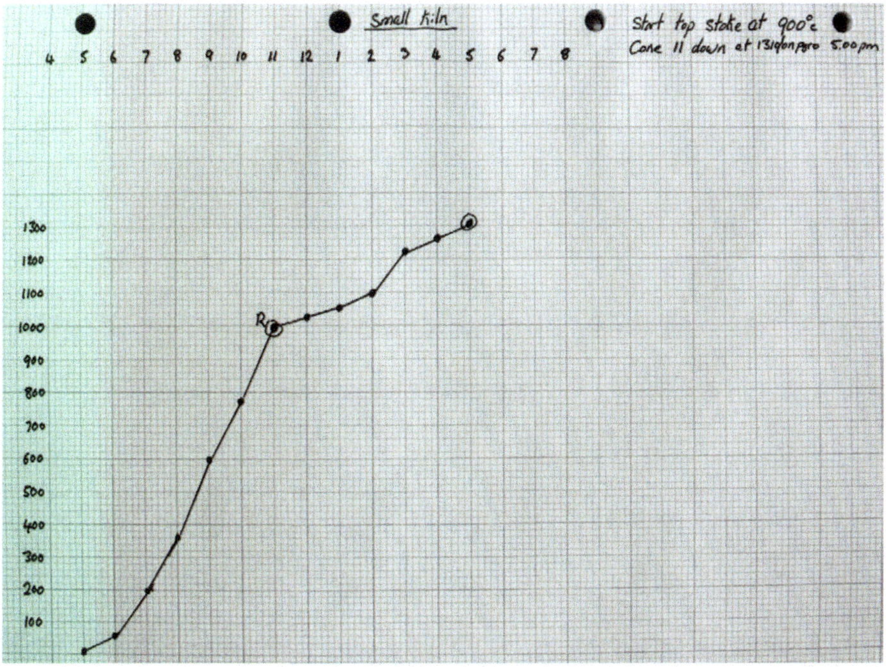

RIGHT: Kiln Graph for Stephen Parry's small wood-fired kiln. R denotes when reduction is started. *Photo: Stephen Parry*

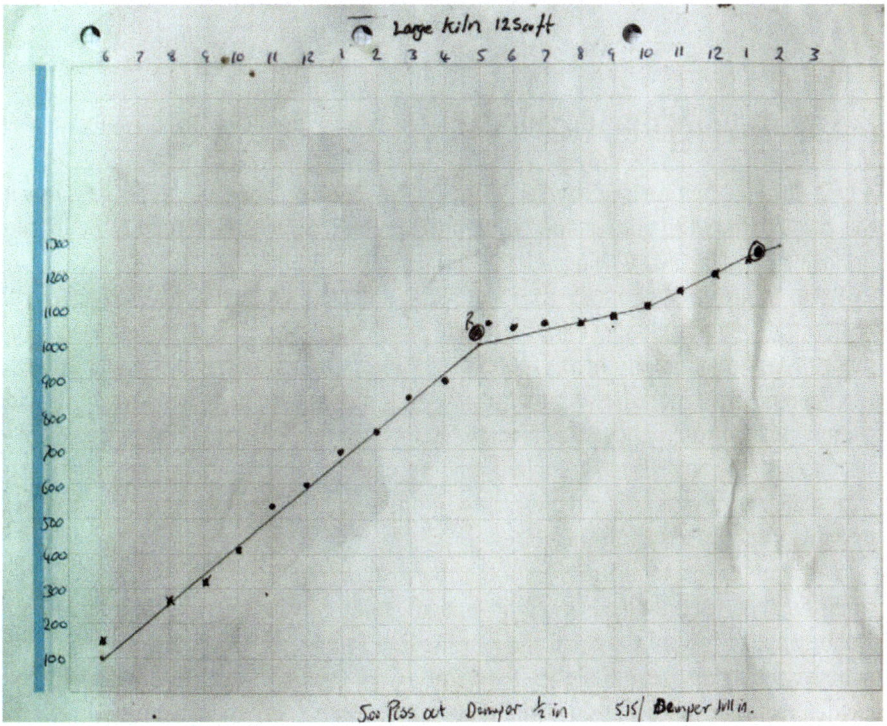

RIGHT: Kiln Graph for Stephen Parry's large Phoenix-type wood-fired kiln. R denotes when reduction is started. *Photo: Stephen Parry*

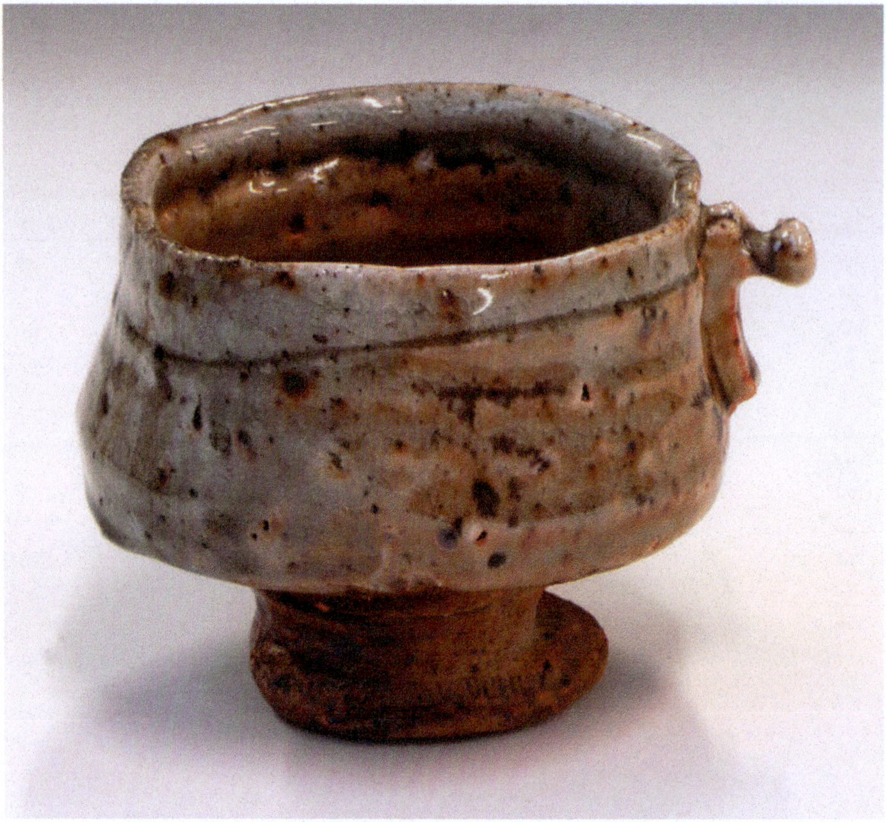

LEFT: *Small Stemmed Bowl,* Benedict Brierley. Thrown and altered wood fired for 4 days in a crossdraft kiln. Height: 12 cm (4¾ in). *Photo: Author*

POST-FIRING REDUCTIONS

A post-firing reduction is one which is applied to a molten glaze, but not necessarily in the kiln in which the glaze is fired. The most frequent applications tend to be in raku firings or on-glaze lustre firings.

Raku firings will be covered in detail in the next chapter, but essentially the post-firing reduction aspect of the firing involves the work being removed from the kiln while at orange heat (1000°C/1832°F), or thereabouts, and being covered with combustible materials. As the glaze is still molten, the action of covering with combustibles instigates those combustibles to try and ignite. If this takes place in a sealed container there is only limited amount of air available for this combustion to take place. Once this is used the carbon monoxide produced liberates oxygen from metallic oxides held within the glaze.

With an on-glaze lustre firing the base glaze will already have been fired, traditionally a 1060/80 lead-based tin-opacified glaze. A paste containing metal salts will be applied to this glaze and the kiln fired up to around 700/800°C (1292/1472°F) in heavy reduction. As with higher-temperature reduction firings, the aim is to liberate oxygen from the metallic oxides and reduce them back to metallic lustres.

RIGHT: *Grand Slam with Shrapnel Lid,* Andrew Niblett. Raku-fired vessel with copper matt lustre, 31 x 23 cm (12 x 9 in). *Photo: Andrew Niblett*

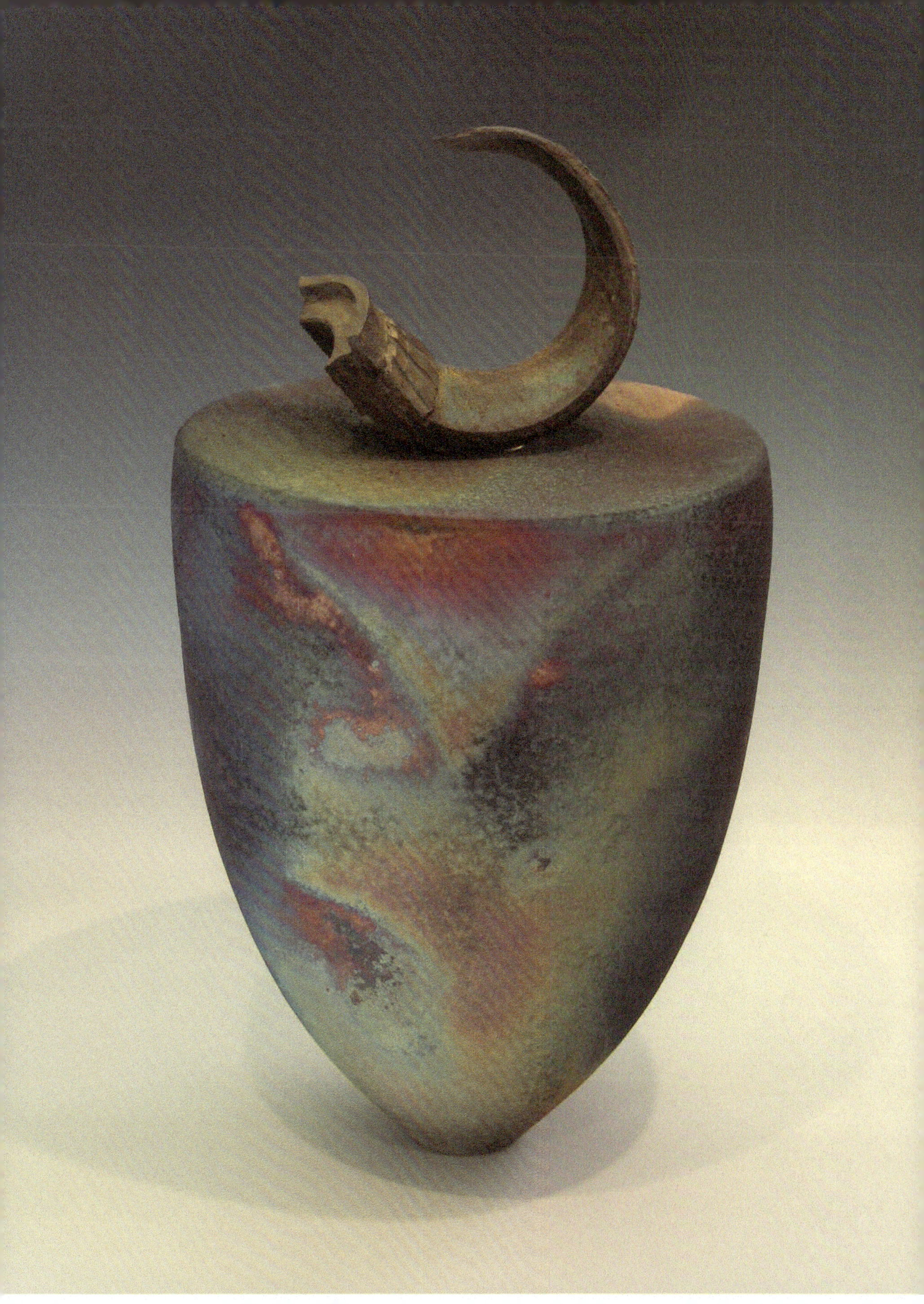

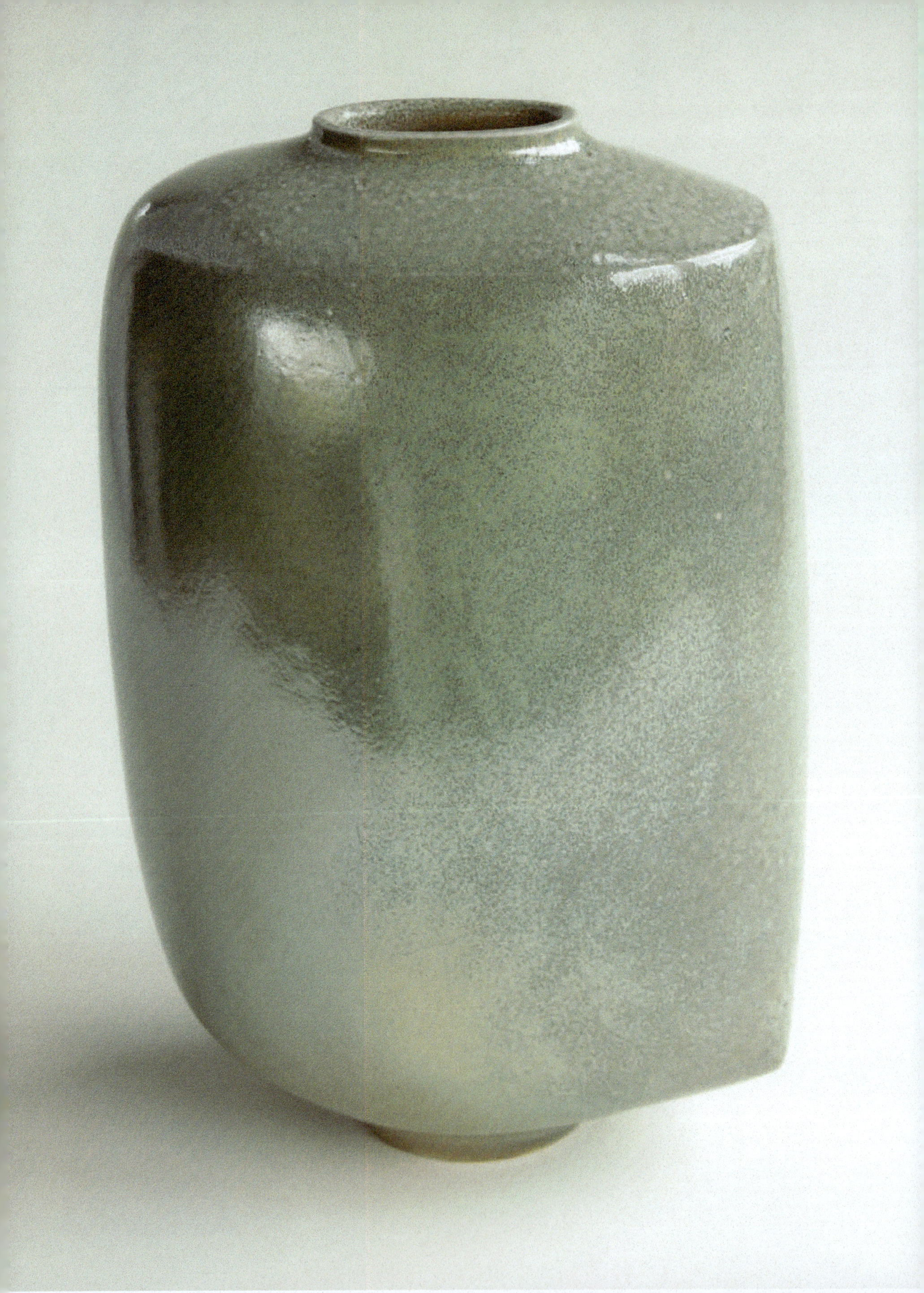

5

Specialist kilns and firings

When considering the firing of ceramics, there can be more to the process than simply generating enough heat to fuse clay or melt a glaze. These firings either rely on the fuel to impart a surface quality, or the addition of another material into the kiln chamber which has a fundamental altering effect on the surface of the work contained within. This chapter explores some of these techniques.

SALT FIRING

Salt firing generally refers to stoneware firings into which salt is introduced in order to develop a sodium silicate glaze. For this to be practical, there are several considerations of both kiln design and construction which should be considered. Salt is very corrosive and will tend to quickly damage soft HTI bricks; therefore it is more usual for kilns which are designed to be fired with salt to be built using heavy, high-alumina fire bricks. However, there are several products available commercially which will offer a degree of protection to HTI bricks and even ceramic fibre. These need to be regularly maintained in order to offer continued protection to the underlying refractory material. Frequency of firing will of course have an influence on how often the protective layer will need to be reapplied. If a kiln is used relatively infrequently then the kiln may go for several years without any work needed.

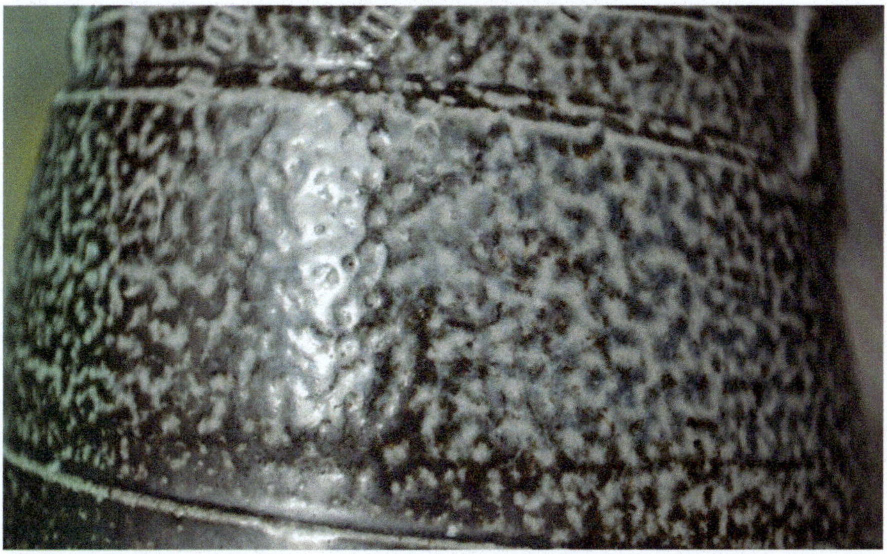

LEFT: *Tall Angled Bottle,* Ruth King. Fired in Ruth's 18 cu. ft propane gas-fired downdraft salt kiln, to cone 9 in reduction, 32 x 22 cm (12½ x 8½ in). *Photo: Peter Heaton*

RIGHT: 'Orange peel' glaze surface. Characteristic of a high-silica slip in salt firing. The slip has been coloured with cobalt and iron. *Photo: Author*

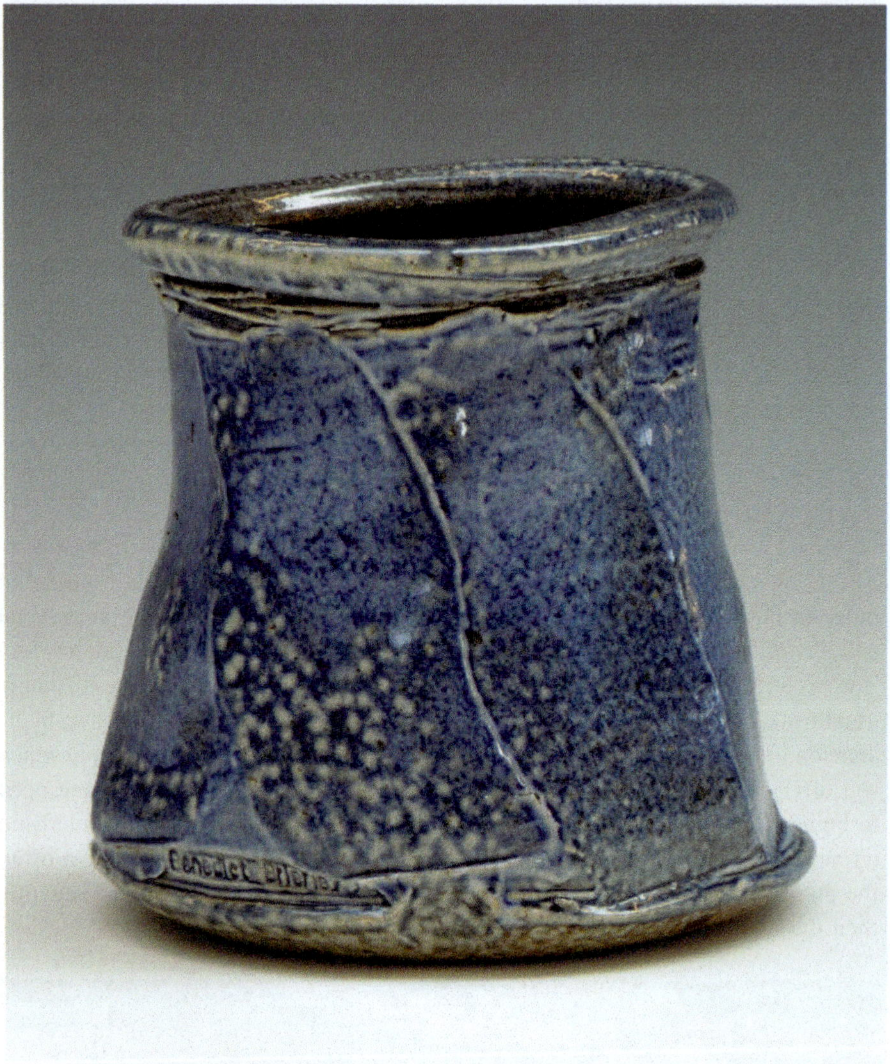

Salt-glazed Vessel, Benedict Brierley. Blue, high-silica salt slip (no. 1 recipe opposite), with orange peel surface, 19 x 13 cm (7½ x 5 in). *Photo: Author*

The active component in a salt firing is sodium. When sodium chloride (common salt) is introduced into the kiln chamber above 1200°C (2192°F), the two elements disassociate. While the chlorine is released from the chimney, the sodium will combine freely with silica and iron in the clay to form a glaze. Salt glaze tends to be characterised by a dimpled surface referred to as 'orange peel', how strong this orange peeling is, depends on the alumina and silica content of either the clay or the salt slip which has been applied.

Alumina does not form a glass with the free sodium so the higher the alumina content of the clay or slip, the least glaze is formed. In reverse, the higher the silica content of the clay or slip, the glassier the surface will become. It is for this reason that often a high-silica slip will be applied to the surface of the work to be fired, which allows for colouring oxides to be added to alter the colour of the final glazed surface.

Salt slip recipes

1. Dark blue

Hyplas ball clay	64
China clay	22
Quartz	14
Cobalt Carbonate	1.5

2. Yellow/orange

AT ball clay	30
China clay	70

The above recipes give examples of a high-silica and therefore glass-forming slip (slip 1) and a higher-alumina and therefore more opaque slip (slip 2).

How salt is introduced into the kiln chamber will depend on kiln design. Gas- or oil-fired kilns will generally have salting ports built into the wall of the kiln above the burner ports, or built into the side wall along the length of the firebox. When the kiln is at the appropriate temperature, salt is introduced in small amounts. When firing salt kilns the author has introduced 3–4 kg (6½–8¾ lbs) of salt into the fireboxes of a 15 cu. ft kiln roughly 500 grams (1 lb) at a time. The salt is wrapped in newspaper in 250-gram batches and then soaked in water. The damp newspaper salt parcels help to keep the salt together as it is introduced into the kiln; additionally, as the water turns to steam and expands it helps to disperse the salt vapours throughout the kiln.

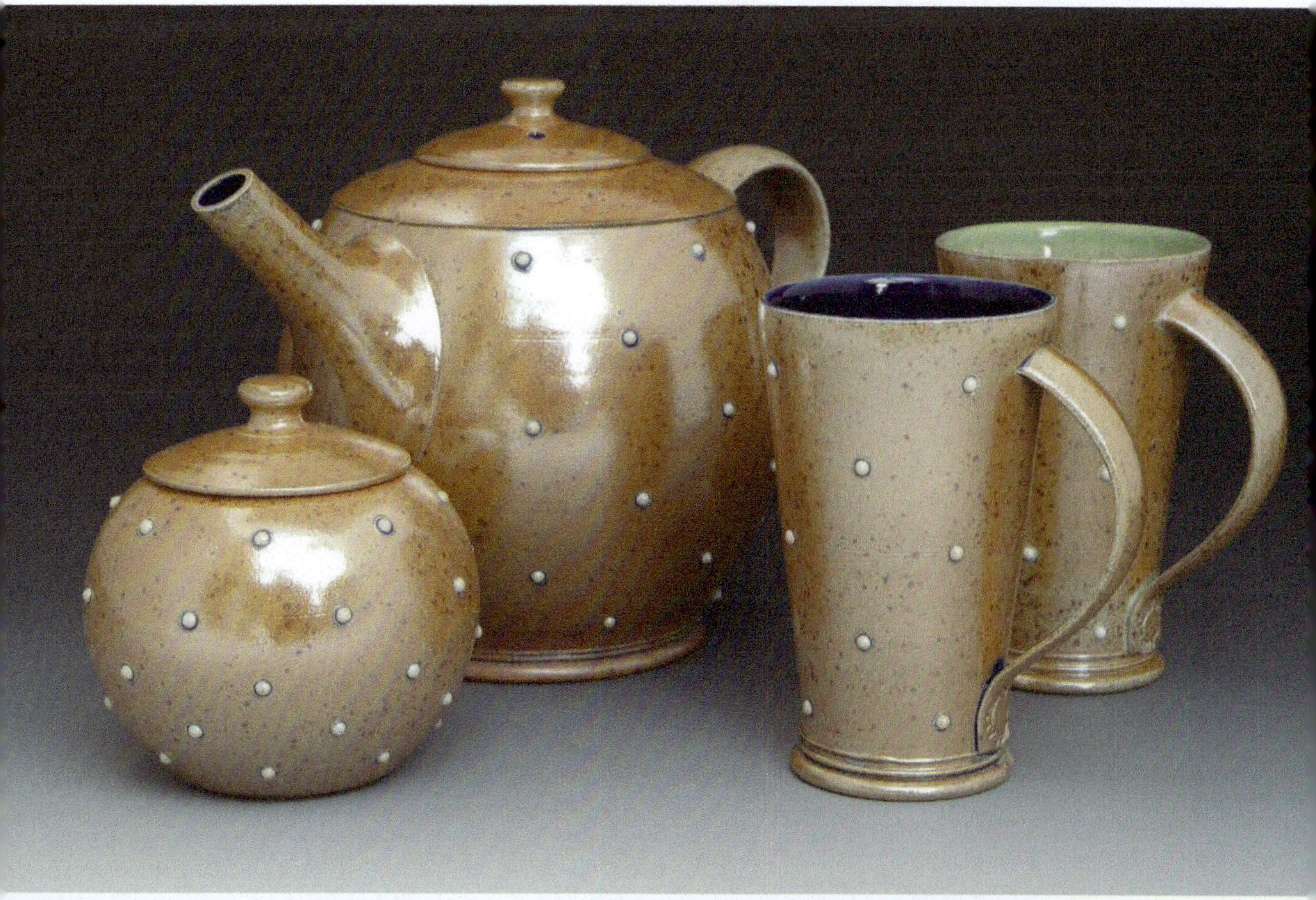

Salt-glazed Tea Set, Alistair Young. Fired in Alistair's gas-fired salt kiln to Orton cone 10 (1280°C/2336°F), Teapot 20 cm (7¾ in). *Photo: Alistair Young*

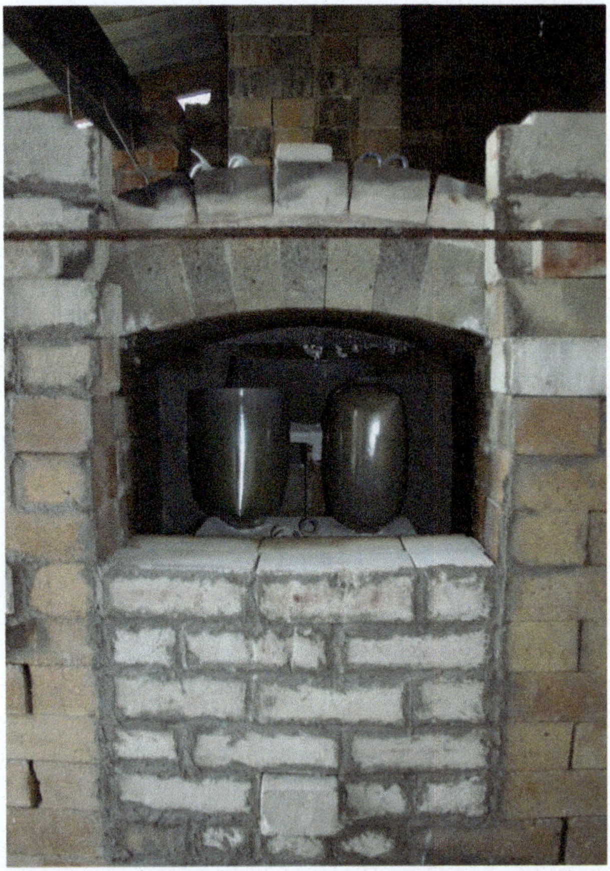

LEFT: A glimpse through the half un-bricked door of Ruth King's downdraft salt kiln. As well as the finished pots, fired cones are visible together with a draw ring. Draw rings are small rings of clay sometimes with a salt slip applied and are commonly used in salt firings when hooked out of the kiln through a spy hole during salting. This gives the maker a good idea of how well the salt glaze is building up on the work. *Photo: Ruth King*

RIGHT: *A Pair of Tall Jugs,* Alistair Young. Salt fired to cone 10 (1280°C/2336°F), height: 81 cm (32 in). Alistair's firing schedule: 'Both burners lit. Oxidised flame until cone 011. Strong reduction using damper and less air for 20 minutes. Steady climb to cone 8 then oxidise for 10 minutes. Salt is then introduced by sprinkling through the side ports using a large stainless steel spoon; a total of 450 grams over one hour. Continue temperature rise until cone 10 is going then gas off, leaving the fans running for 20–30 minutes until the temperature is below 1000°C (1832°F) then burners are unbolted and blocked off. Damper in.' Total firing time 12–13 hours. *Photo: Alistair Young*

Salt packages are introduced into the kiln chamber by means of a piece of angle iron, or laid out on a piece of wood. The salt parcel is laid onto the angle iron and the end of the iron placed into the salting port, the salt parcel is then pushed into the salt port by means of a piece of metal rod. Salt is introduced into the flame path of the burners so that the flame also helps to disperse salt around the kiln more efficiently. In wood-fired salt kilns the wetted salt packs are often simply placed on a piece of wood and placed into the firebox; as the wood combusts, the salt is carried around the kiln.

Salt vapour is indiscriminate and will glaze everything which contains silica; therefore, the whole of the interior of salt kilns will eventually become glazed. In order for work to be released from the kiln after firing, each piece must be placed on wadding. Salt-kiln wadding is made of high-alumina materials to resist the salt vapour and will tend to leave a small resisted area on the bottom of the work. This application of wadding can be used decoratively. Seashells may also be used to set the work off the shelf, usually reinforced with wadding. Shells will leave a decorative mark in the salt glaze and will also dissolve when wetted after the firing, enabling work to be easily released. There are many different recipes for wadding, however two examples are included here for reference.

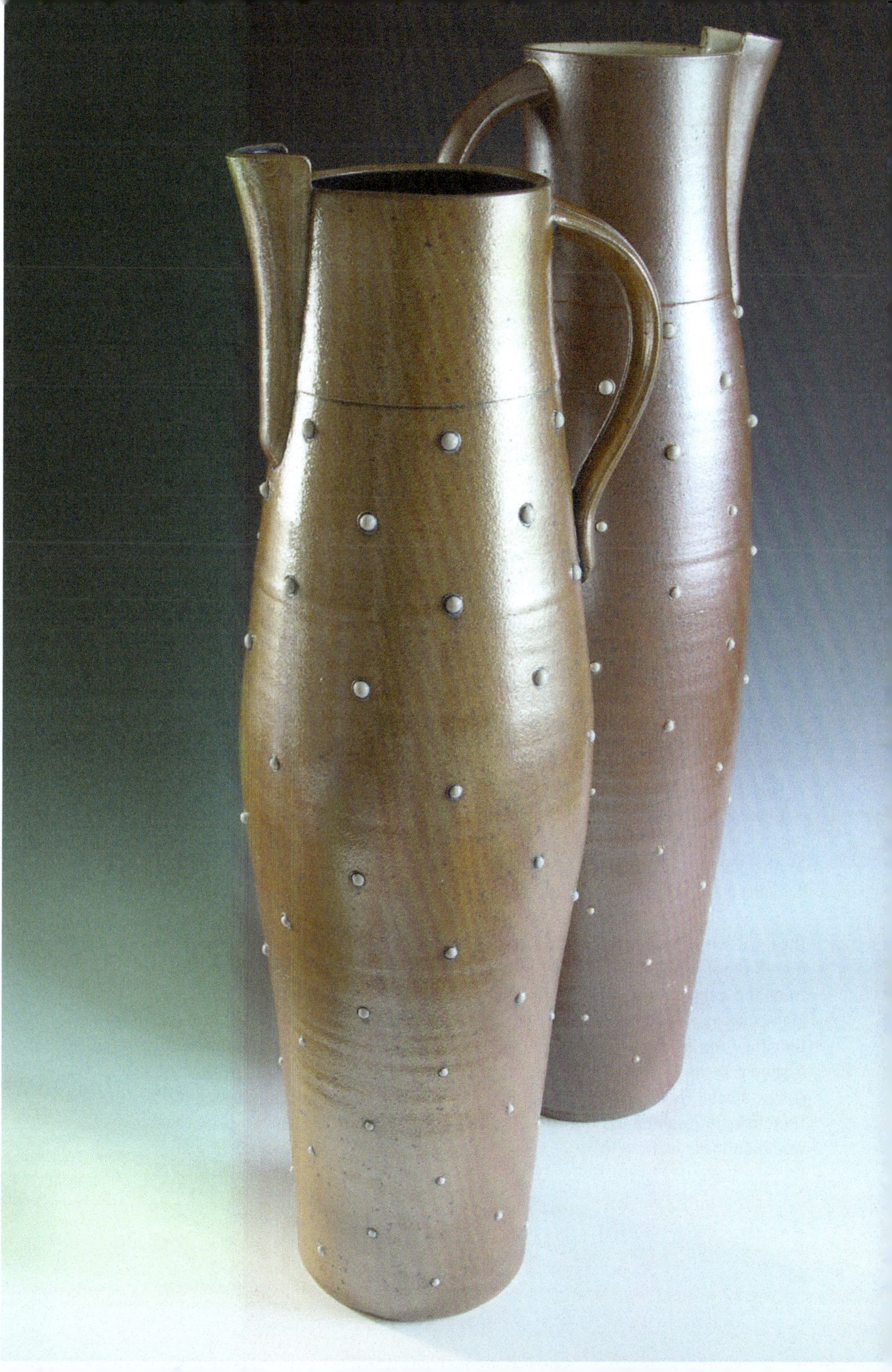

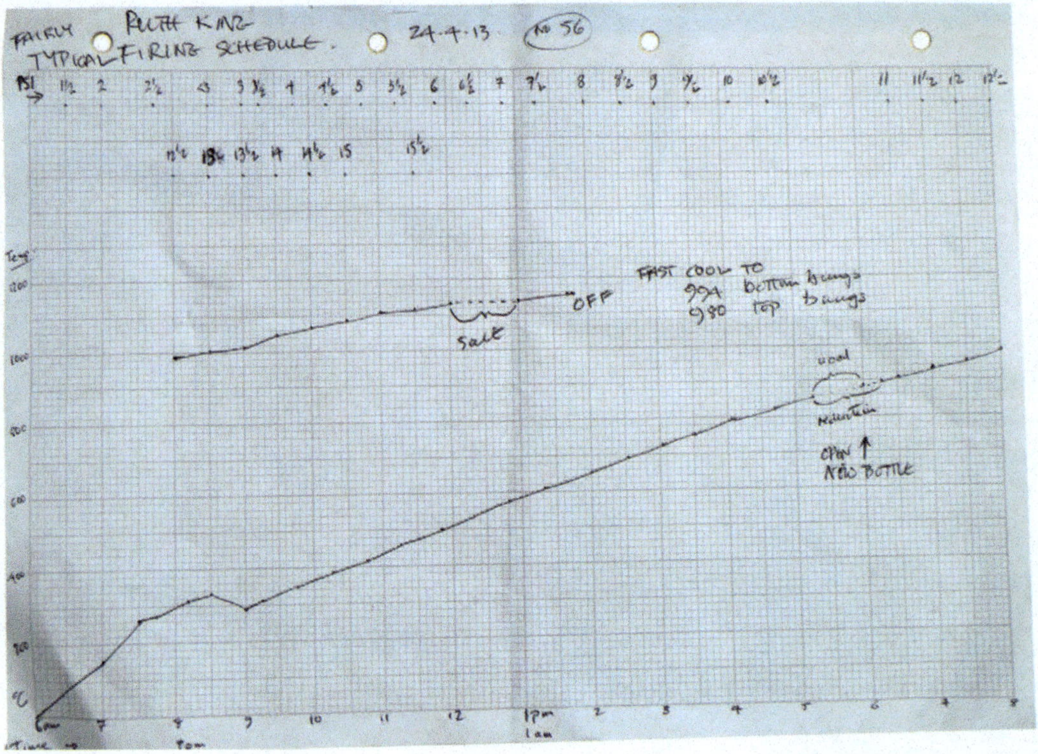

ABOVE: Firing schedule used by Ruth King in her 20-hour salt firings. *Photo: Ruth King*

Salt kiln wadding recipes: (by weight)

1. Alumina 6
Flour 6
China clay 1

2. Alumina 5
China clay 5

The inclusion of combustible materials helps to make the wadding more crumbly and therefore easier to remove. After the salt is introduced, the kiln emits white smoke from the chimney and when this smoke clears the next salting can commence. Many makers when firing with salt position several draw rings close to spy holes so that they can be removed after each salting and the salt glaze build-up assessed. It should be noted that the sodium which forms the glaze also causes pyrometric cones to bend earlier than they would do normally, due to soda being a strong flux. Therefore, it is sometimes useful to put a higher guard cone into the pack than would normally be fired to.

RIGHT: *Deep angled bowl*, Ruth King. Salt glazed, 28 x 26 cm (11 x 10 in). *Photo: Peter Heaton*

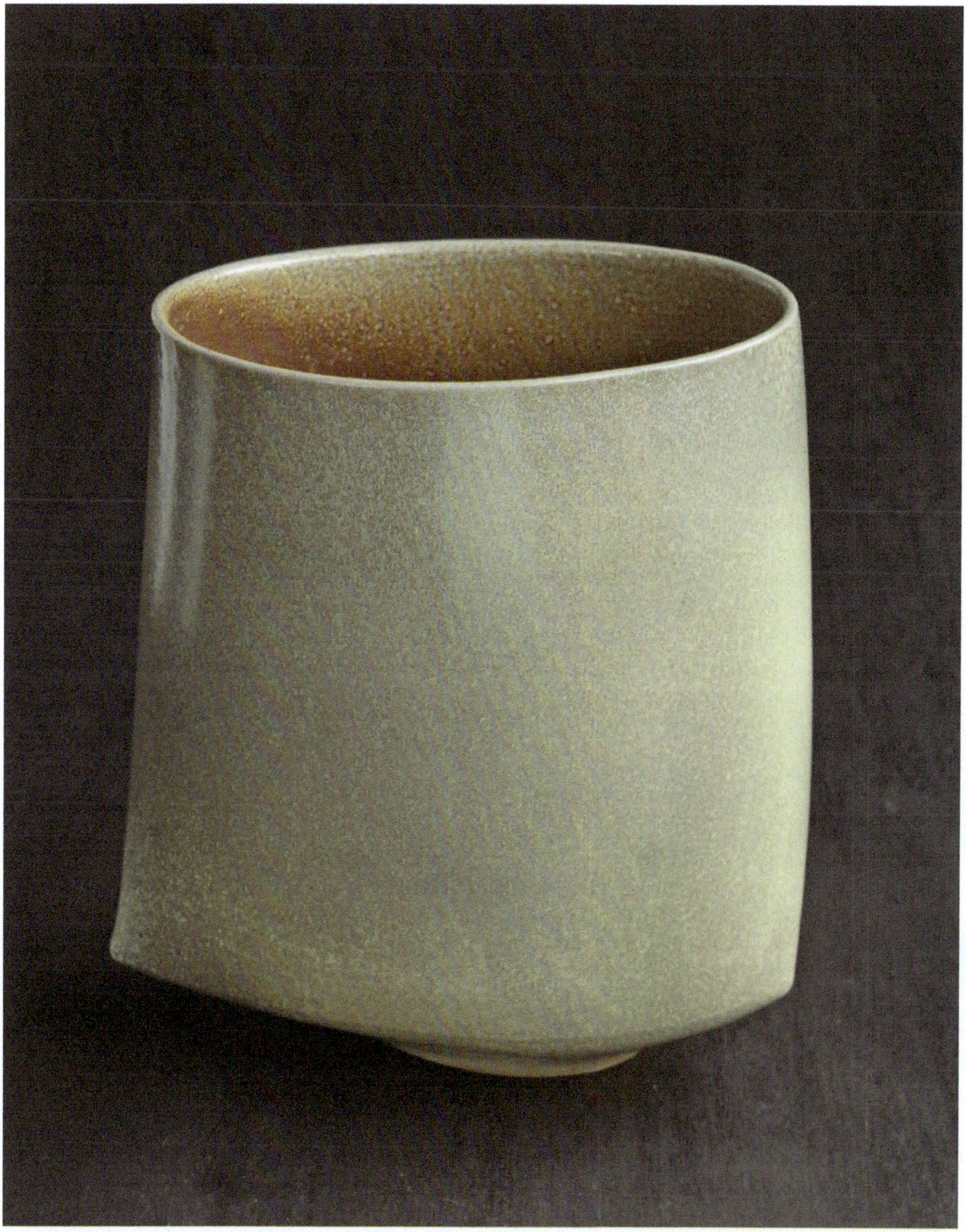

SODA FIRING

RIGHT: *Wood- and Soda-fired Porcelain Tumbler,* Thomas Aitken. 12.5 x 8.5 cm (5 x 3¾ in). This piece demonstrates beautifully the combination of body flashing and soda orange peel achievable in this combination firing. Soda surface tends to look 'softer' than a salt-glazed surface. *Photo: Thomas Aitken*

Soda firing in many ways is similar to that of salt firing, in that it is sodium which is the active glaze component. However, there are specific differences in the way in which soda behaves in the kiln. Sodium in soda firing is usually introduced in the form of soda ash (sodium carbonate) or bicarbonate of soda. Sodium in this form will not travel as well around the kiln as salt, and so its introduction tends to be more direct and focused. Generally the soda ash will be dissolved in hot water and a solution made of soda-saturated water (although with wood kilns it can be mixed with water to a paste and smeared onto wood in much the same way as salt). This mixture is then sprayed into the kiln. The solution can be sprayed into the fireboxes in the same way that salt is introduced, however due to its reluctance to travel through the kiln well it can also be sprayed directly on to the work inside the chamber. To facilitate this, kilns designed for soda firing will tend to have several soda ports built into the walls on either side, this will allow soda to be sprayed into the whole kiln pack. Because soda can be sprayed, it allows for decorative use in that selected surfaces can have a heavier application which can give dramatic faces and surfaces to the work.

In both soda and salt firings the soda vapours will struggle to get inside more enclosed forms, so it is usual for a liner glaze to be applied to the interiors, while a slip is applied to exteriors. As with salt, work being fired in a soda kiln needs to be wadded to avoid sticking to kiln shelves, etc. Kiln shelves used in this type of firing should have a layer of batt wash applied to the top surface of the shelf; this also aids the release of work after the firing.

Batt wash for salt/soda firings

Alumina	50
China clay	50

Wash can also be applied to the tops of any kiln props used to stop shelf and props fusing. It is not, however, advisable to wash the bottoms of shelves as soda can get beneath the layer of wash and cause it to fall onto the work below.

When firing salt or soda kilns it is down to personal preference as to the schedule of the firing and at what temperature soda is added, but this tends to be from 1200°C (2192°F) and above. However, there are several schools of thought as to how a firing should be ended. Generally a firing will be carried out with predominantly a reduced atmosphere; this can be fluctuated slightly as use of the damper can help to keep the volatilised sodium vapour in the chamber for longer, increasing its association with body and slip silica. Some makers argue that oxidising the kiln for a short time at the end of the firing brightens up the glaze, while others will crash cool the kiln at the end of a firing to harden and brighten the glaze.

Generally speaking, cone packs for salt and soda firing will contain a 06/07 (980–1000°C (1796–1832°F) Orton), to indicate the start of reduction followed by a cone 6 (1220°C/2228°F) or cone 7 (1240°C/2264°F) to indicate the start of salting. This

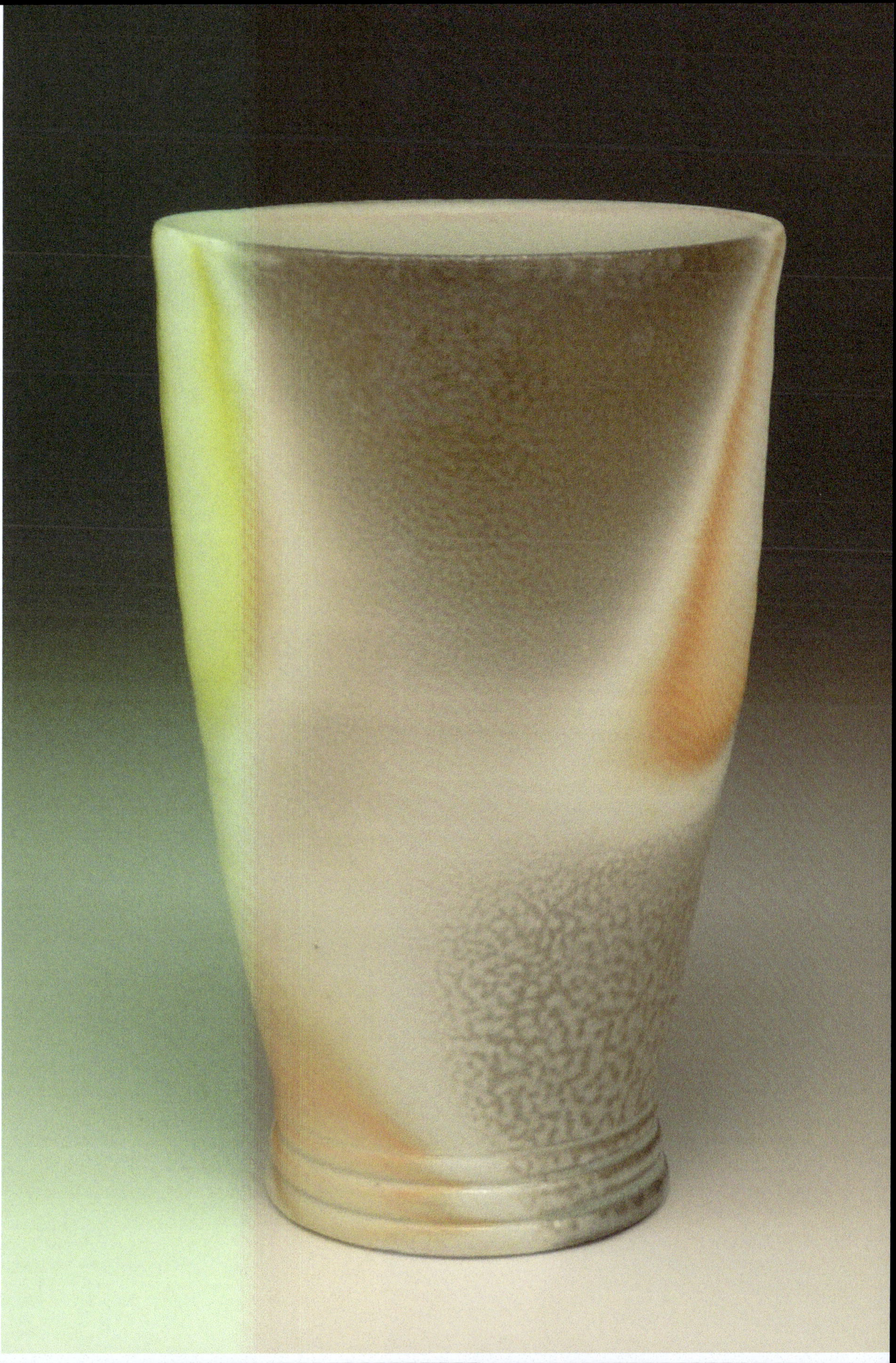

second temperature is not fixed; some makers choose to start salting earlier or later depending on preference. At 1200°C (2192°F) the clay surface is starting to get a little softer and so this provides a good key for the soda to bond. A good reduction through the firing also reduces the iron which becomes a strong flux adding to the surface qualities. The next cone in the pack could be 8 (1250°C/2282°F), 9 (1260°C/2300°F), 10 (1280°C/2336°F) or 11 (1300°C/2372°F), depending again on preference. As was mentioned earlier sodium from the salt or soda will cause cones to bend slightly earlier and it could be anticipated that a cone 11 bending would not be far off a cone 10 fired in a gas kiln.

RIGHT: *Low-temp salt-fired vessel,* Oscar Copping. This piece was tumble stacked in a bespoke kiln surrounded by small packets of salt and fired to approximately 1100°C (2012°F). Heavily reduced from 1000°C (1832°F) and soaked for 3 hours at top temperature, 25 x 22 cm (9¾ x 8½ in). *Photo: Author*

Firing schedule for 2-burner gas-fired salt-kiln firing

9am	2 x pilot burners lit
10.30am	2 x main burners lit
11.30am	All burners turned up
12.30pm	Cone 06 bending. Increase burners and insert damper to start medium reduction
3pm	Cone 6 bending; start salting. Push in damper a little to slightly increase reduction
3.15pm	2 x 250g (9oz) salt packs in (one each firebox)
3.30pm	2 x 250g salt packs in (one each firebox)
3.45pm	2 x 250g salt packs in (one each firebox)
4pm	2 x 250g salt packs in (one each firebox) Cone 9 bending
4.15pm	2 x 250g salt packs in (one each firebox)
4.30pm	2 x 250g salt packs in (one each firebox)
4.45pm	2 x 250g salt packs in (one each firebox)
5pm	2 x 250g salt packs in (one each firebox)
5.15pm	Cone 10 bending
5.45pm	Cone 10 flat. Turn all burners off, open damper and all secondary air and spy holes to crash cool
6.15pm	Close damper, secondary air and spy holes and allow to cool.

Finish

OPPOSITE, LEFT TO RIGHT:

A bespoke low-salt kiln being tumble stacked by students at Loughborough University School of the Arts.

Low-salt kiln firing. The kiln is fired using a swirlamiser oil burner.

Low-salt kiln being unpacked.

Photos: Author

FUME FIRING (LOW-TEMP SALT)

A low-temperature salt or fume firing varies from high-temperature salt firing not only in temperature, but also in the way that salt is introduced into the kiln, the pack and how the salt reacts with the ceramic.

A kiln which is to be used for low salting and fuming should be constructed out of high-alumina firebrick. Kiln furniture for this type of firing should be reserved just for low-salt firings, this is due to salt vapour tending to penetrate refractory shelves and brick work. If the kiln is then fired high, the accumulation of salt may cause shelves

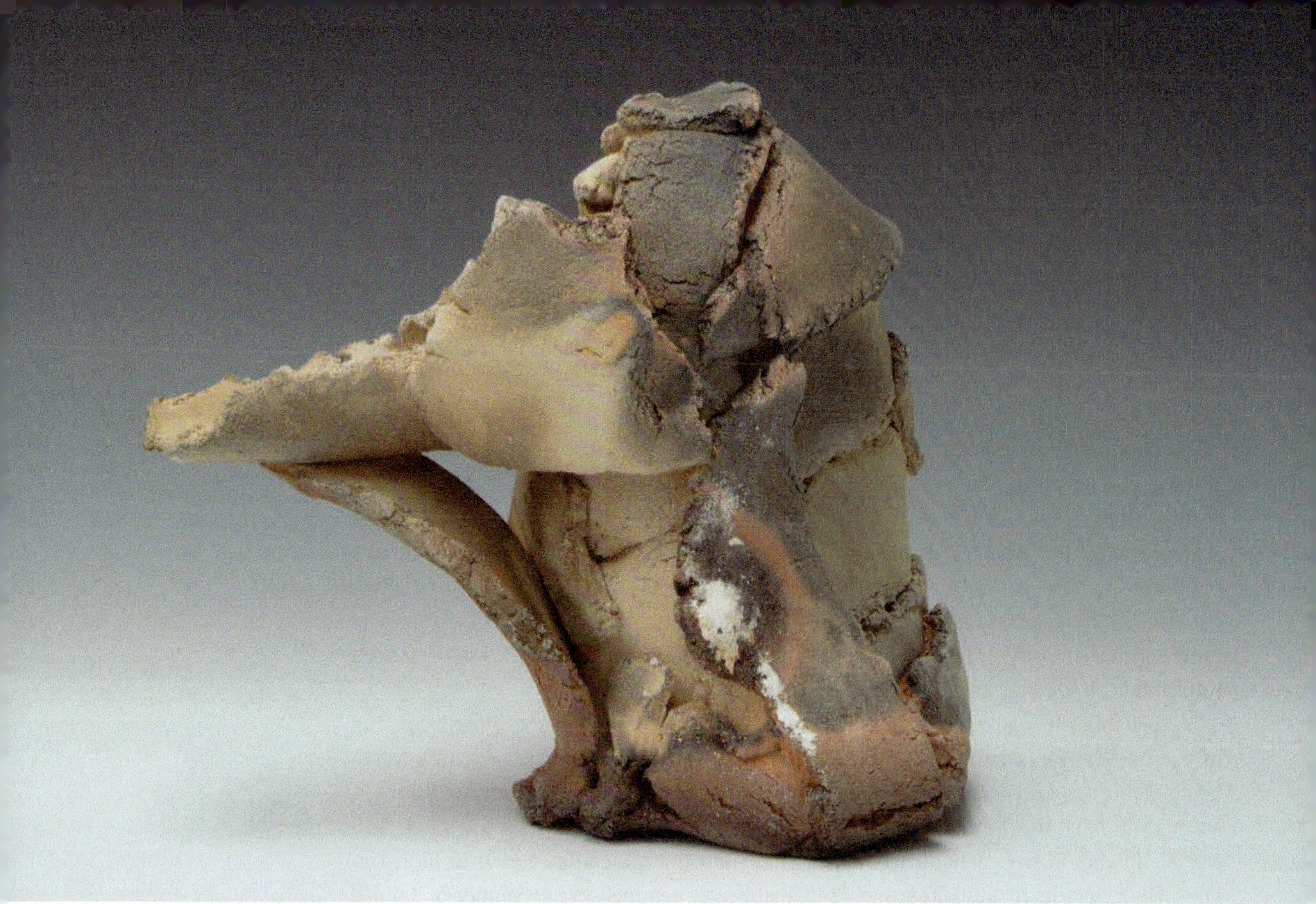

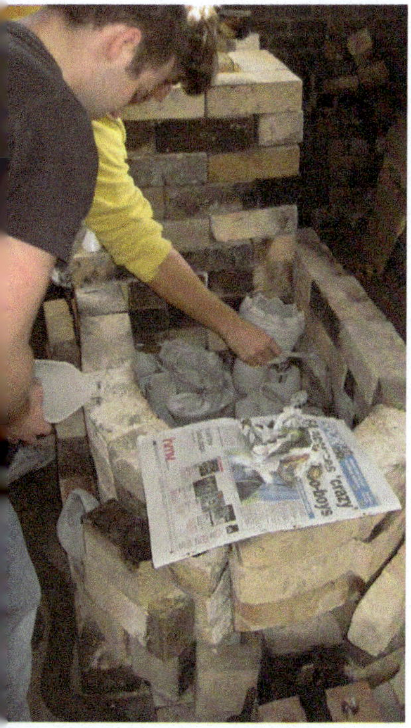

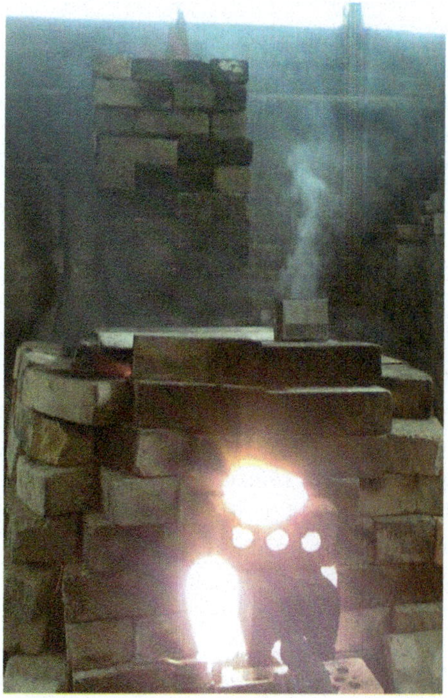

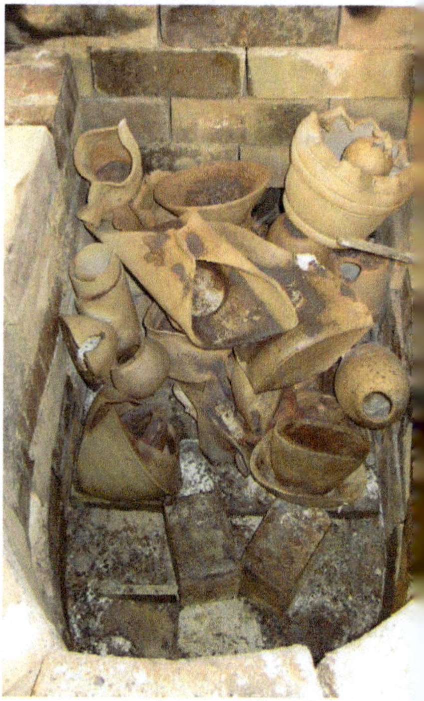

to fail. If washed with the same bat wash mix as for high salt and used specifically for fume firings they should last a good number of firings.

In a low-salt firing, the kiln is heated to around 1000°C (1832°F) and then reduced very heavily. Reducing heavily at this temperature creates a richer clay surface colour as it is just at the bottom end of the point at which iron starts to reduce and flux, it also holds the salt fumes in the chamber for longer, forcing flame paths through the work which increases reaction with the clay body. The temperature, at which the kiln is reduced, is not high enough for the salt vapour to volatilise and form a glaze, so the surfaces achievable with this type of firing tend to be matt in nature.

A common surface treatment in preparation for this type of firing is terra sigillata. Terra sigillata is a slip made from the very finest clay particles. Powdered clay is mixed with plenty of water and a small amount of deflocculant. The blend is mixed well and then allowed to settle, the heaviest particles fall to the bottom and the finest are held in suspension. The top layer which is usually water is removed and the middle layer containing the fine suspended particles is then decanted off and brushed or sprayed onto leatherhard, dry or biscuit-fired clay. A terra sigillata should have a warm sheen to it and is an ideal surface to take low-temperature fuming.

A low-salt kiln pack tends to be tumble stacked. Pots are placed directly on top of each other with no shelves, and packages of salt placed between them. It is worth using well grogged clays; certainly for work to be placed at the bottom of the stack, as this will have to support the loading of subsequent pots further up the stack. Salt packs can vary in size, generally being smaller than the quantities added to high-salt kilns. Pack weights of 50–100 grams (2–3½ ounces), wrapped in newspaper or similar, and sandwiched between pots should be ample. Larger work tends to be placed at the bottom of the stack with the work getting progressively smaller, so that the most delicate pieces are at the top.

During firing, the size of the salt packs will decrease as paper containing salt combusts and the salt volume diminishes, therefore care should be taken to ensure that the tumble stacked pots will remain stable. This may be achieved by setting bricks at the base of the kiln to hold the bottom pots in place so that they can't roll and destabilise the stack above them. As with any tumble stacking in a kiln, there is always a chance of pieces moving more than was expected, but the resulting work can sometimes produce the most striking surfaces. Small pots containing salt may be placed around the work directly in contact with the shelf. Wadding does not need to be used as the salt will not volatilise, however wadding may be used to mask areas of work to resist the fuming effect of the salt. Due to the low temperature of the firing, work will still be porous so it is chiefly used for decorative effect on more sculptural pieces.

Any fuel may be used for heating the kiln, apart from electricity due to the corrosive effect of the salt vapour on the kiln elements and the HTI brickwork. The author has used both oil and LPG for firing low-salt kilns to good effect in crossdraft, downdraft and updraft kilns.

RAKU FIRING

Raku firing may be carried out in any fuel-burning kiln in a similar way to low-salt firing; however, the kiln should allow access to the work for removal while the kiln is hot. What tends to be referred to now as raku is a technique which was developed in the 1960s and was derived from an older Japanese technique used in the production of predominantly tea ceremony ware. The modern technique involves heating up biscuit-fired work fast, to around 1000°C (1832°F). Once the kiln has reached temperature it is removed from the kiln and either allowed to cool naturally or placed into a container holding some kind of combustible material such as newspaper, sawdust, leaves, etc.

Low-temperature glazes may be applied to ware being raku fired. These glazes are usually composed mainly of frit, commonly alkaline or borax based, but not lead as this will not tolerate any reduction as mentioned in chapter 3. To these low-temperature glaze bases, colouring oxides or stains may be added. Copper oxide or carbonate can give arguably the most dramatic results. The technique of plunging a still red-hot piece of work into a combustible material induces a post-firing reduction which affects ceramic oxides in a similar way to reduction in a high-temperature firing. Copper, when fired in an oxidised atmosphere in an alkaline glaze base, will tend to give a rich turquoise colour; however, when subjected to a post-firing reduction, and as long as the pot is placed quickly from the kiln into the combustible material, either copper reds or bright metallic surfaces may be achieved as the copper oxide reduces to copper.

Kilns designed for raku firings must fulfil several criteria not really applicable to any other firing, because the work needs to be removed while the kiln is still at orange heat. There are several types of raku kiln design which facilitate this. A wire-framed top-hat kiln lined with ceramic fibre is probably the simplest design.

BELOW LEFT: Small propane gas-fired raku kiln. *Photo: John Wheeldon*

RIGHT: Flat-pack propane gas-fired raku kiln. *Photo: John Wheeldon*

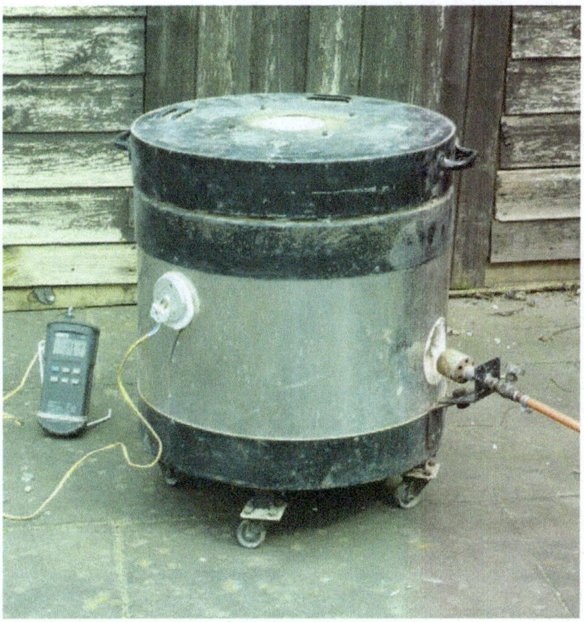

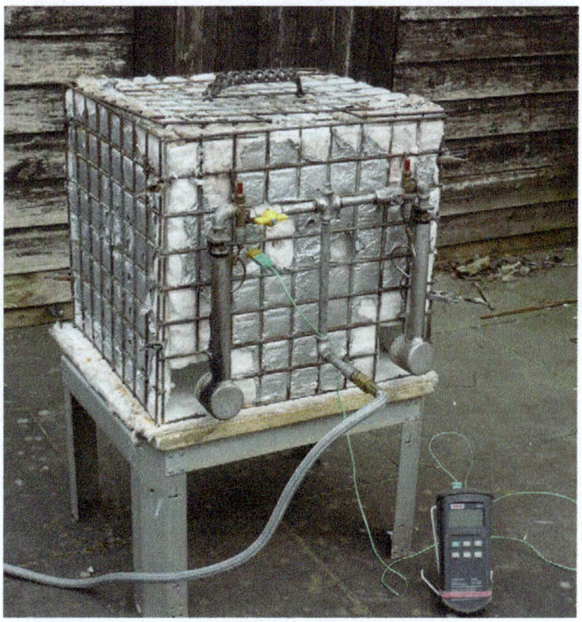

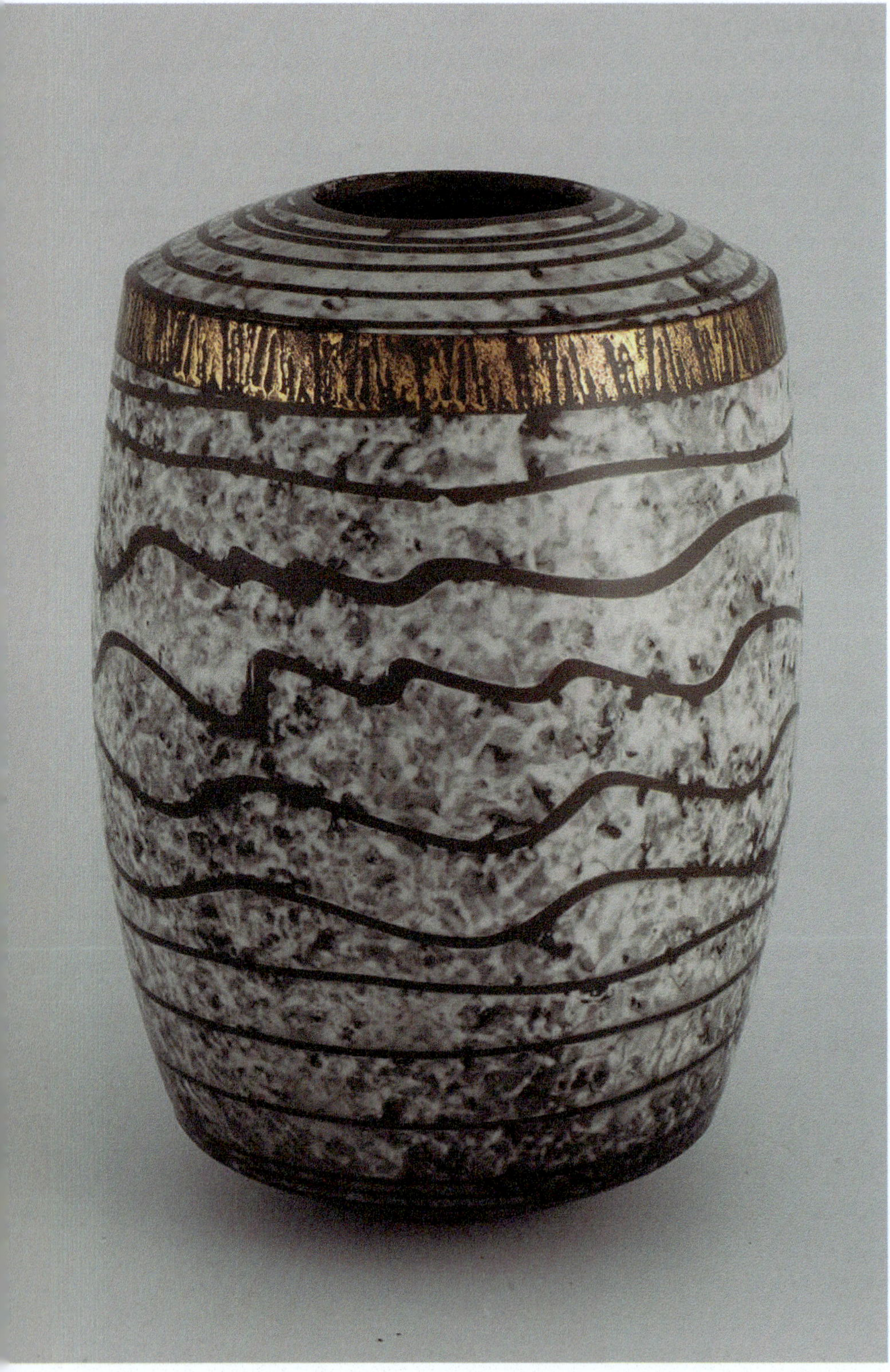

LEFT: *Raku-fired terra sigillata vessel,* with lustre decoration, John Wheeldon. Lustre fired to 650°C (1202°F) in propane raku kiln, 16 cm (6½ in). *Photo: John Wheeldon*

RIGHT: *Copper matt vessel,* John Wheeldon. Height: 18 cm (7 in). *Photo: John Wheeldon*

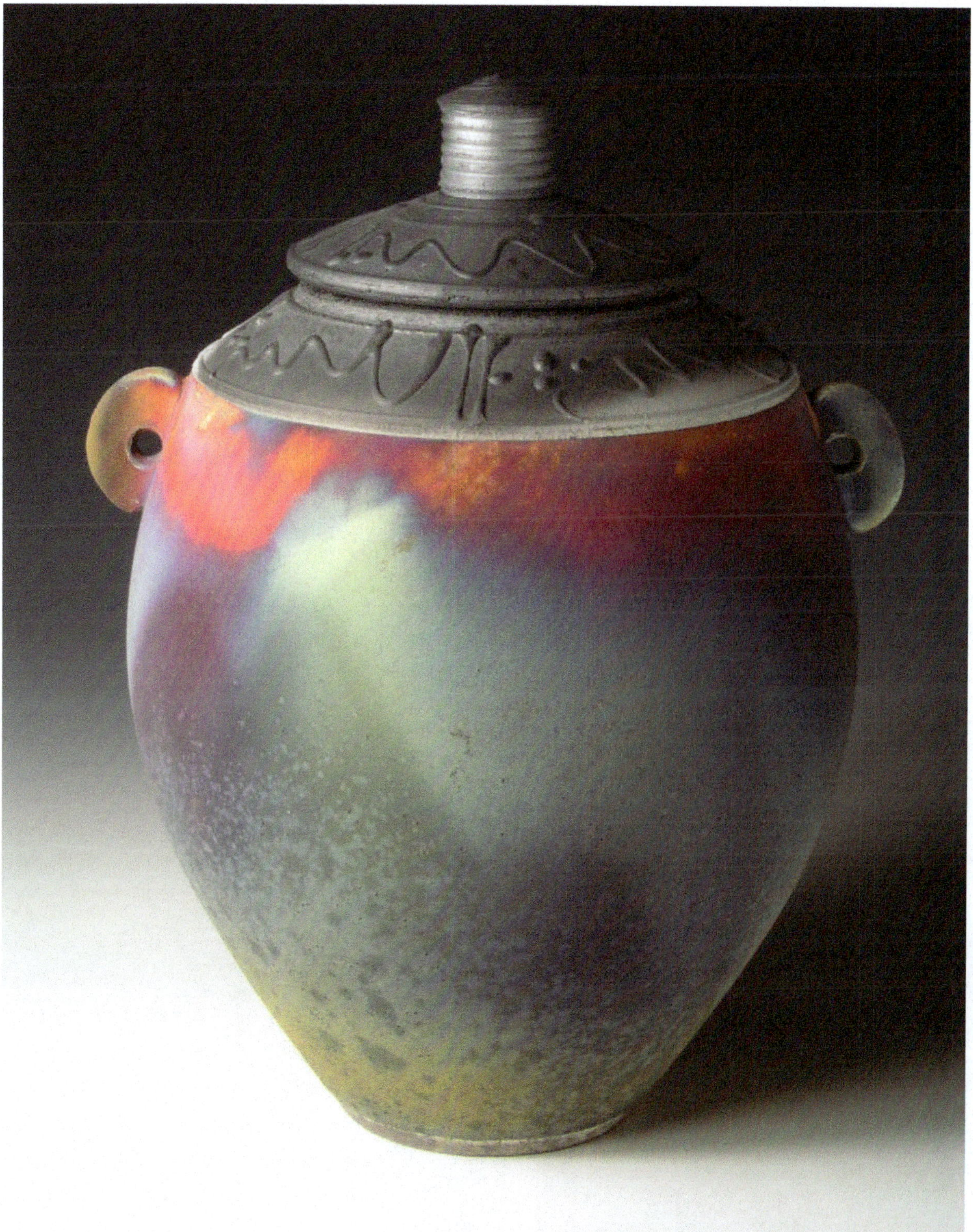

A foundation of brick is used which may be two or three courses high either circular or square with a break in the wall for the burner to go through. If oil is to be used then the quarl will sit in the hole. For a raku kiln it is possible to use domestic house bricks as these are generally capable of tolerating heats up to around 1060°C (1940°F), however, if firebricks are available then these would be preferable. Three bricks are then placed in the centre of the foundations and a kiln shelf placed on these. The flame path from the burner should fire under the shelf although if the foundation aperture is angled, the flame could be fired around the wall – but this only really works in a circular structure. The ceramic fibre 'top hat' can then be positioned on the foundation bricks.

Once the kiln is packed and fired up to the desired temperature, the lightweight fibre lid may be lifted, exposing the work which can then be removed with tongs and placed into combustible materials, or placed in water to fast cool. Old front-loading electric kilns may be converted for use as raku kilns. Simply by removing the elements, cutting a hole in the back wall for a burner to be placed, and a slot cut through the back of the top of the kiln, to allow gases to escape. Another common raku kiln design is again a circular kiln with a fibre lid. When the kiln is up to temperature the lid is removed and the work extracted, as illustrated by John Wheeldon's kiln.

PIT FIRING

A pit firing arguably is the closest form of firing which replicates the earliest kilns. Work is placed on a bed of sawdust at the bottom of a pit, usually dug into the ground although it could be built out of brick. The work is then covered with fuel in the form of sticks and logs and sawdust. To this mix, metal oxides may be added to promote flashing on the surface of the work. The pit is then ignited and covered with a lid such as a sheet of corrugated iron or large earth sods and allowed to burn down.

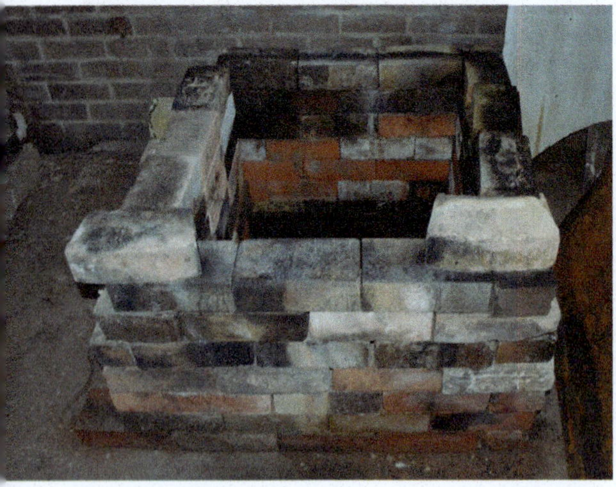

BELOW LEFT: A small smoke-firing pit built from firebricks. House bricks are just as suitable or a metal dustbin with a few air holes drilled towards the base. *Photo: Author*

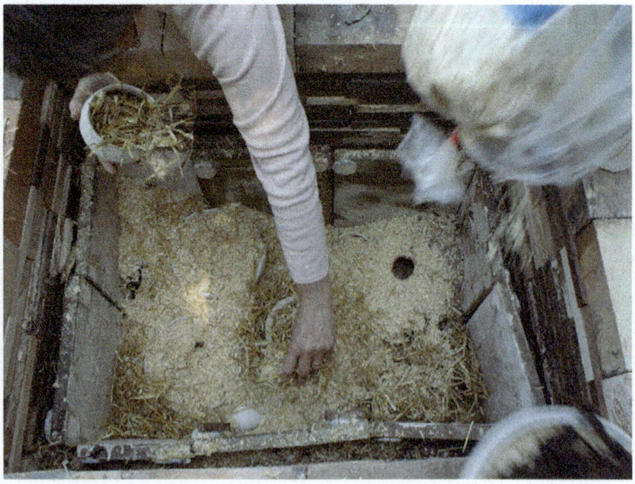

BELOW RIGHT: Beginning to pack a pit firing. *Photo: Lynne McGechie*

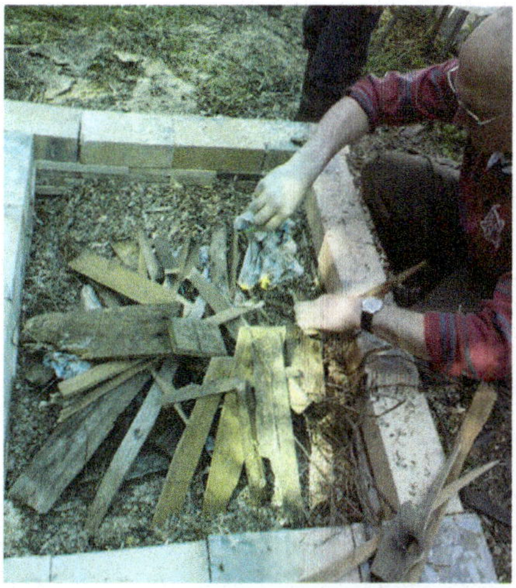

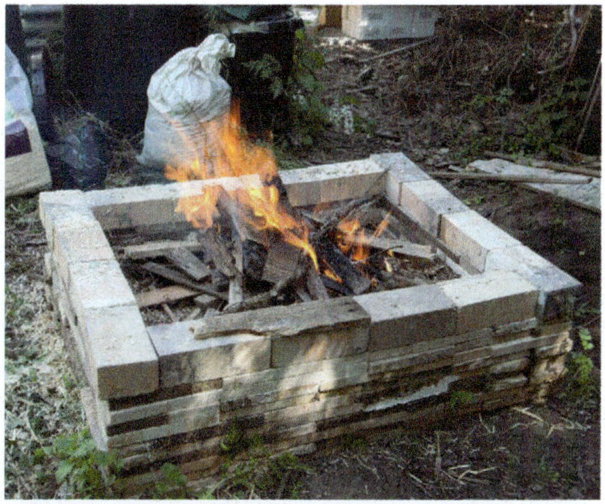

ABOVE LEFT: Lighting the pit.

ABOVE RIGHT: The pit fire taking hold.

RIGHT: Larger pieces of wood are added to the pit to build temperature.

Photos: Lynne McGechie

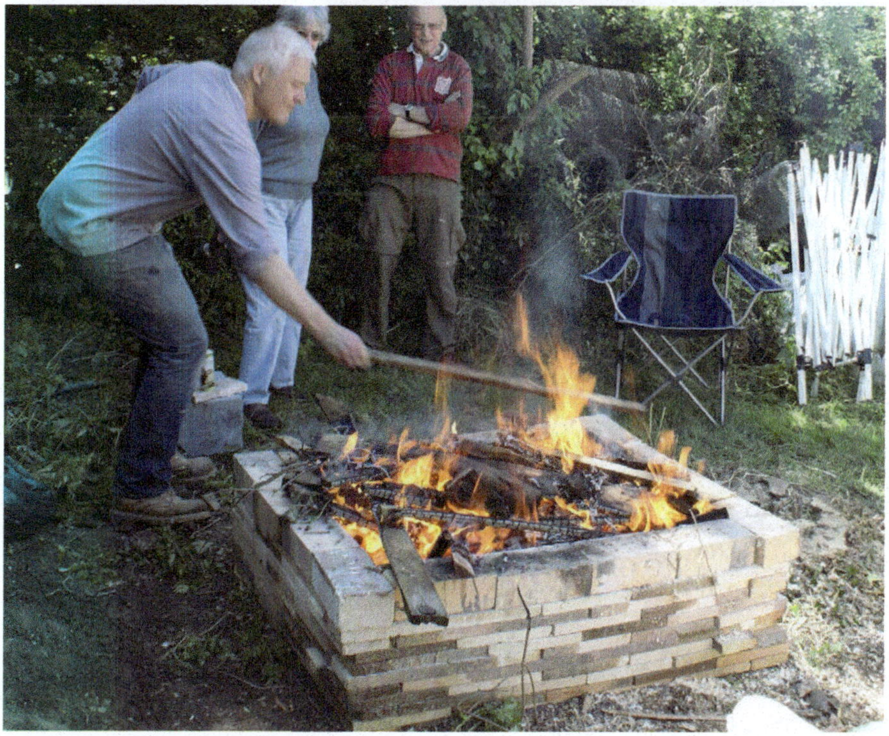

As with raku firing and low-salt firing, robust clay bodies capable of withstanding stresses and thermal shock inherent in these types of firings are beneficial; crank or clays designed specifically for raku-type firings are ideal. Work would normally be biscuit fired before being placed into the pit, the surface of work to be fired in this way could also be treated with a terra sigillata. Today (in European countries) the work is often biscuit fired in an electric kiln first to avoid too many breakages, while the pit firing is used mainly to create surface effects.

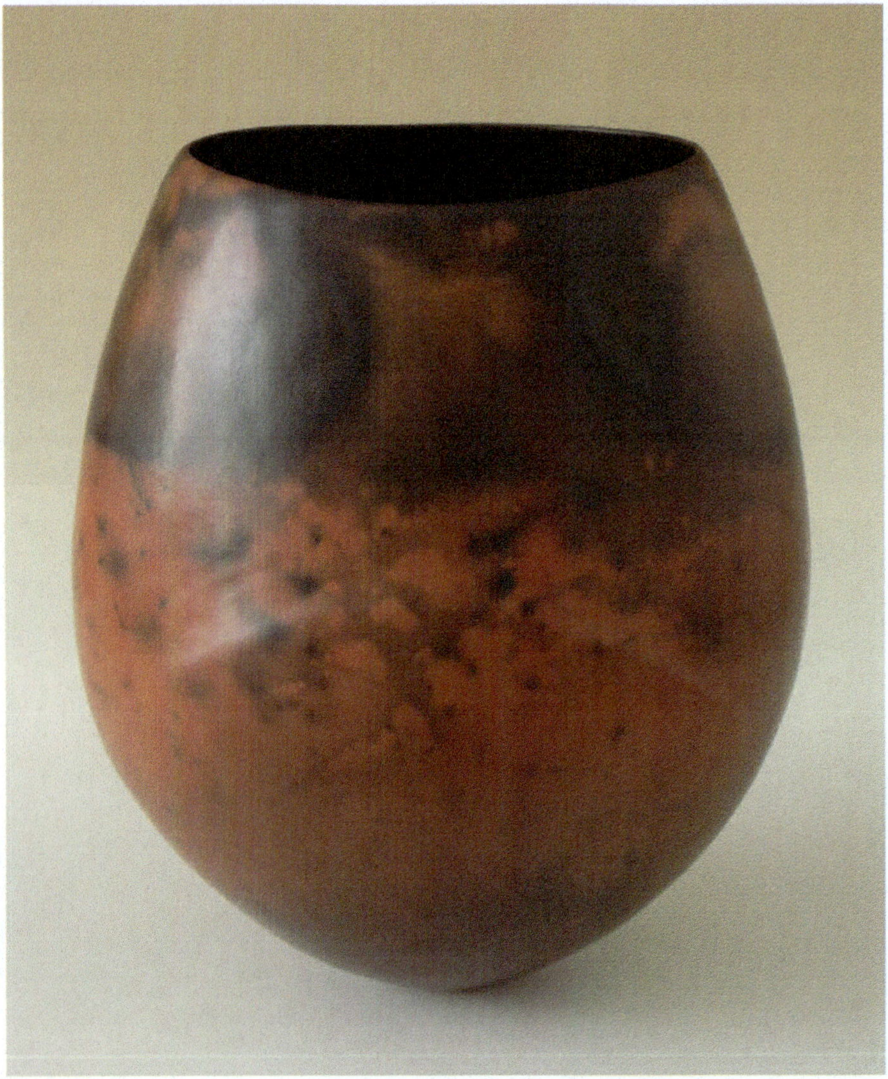

LEFT: *Red Bell*, Gabriele Koch. Porcelain enriched stoneware clay. Smoke fired, 39 cm (15½ in). *Photo: Kelpie*

RIGHT: *Orkney Stone*, Gabriele Koch. Height: 58 cm (23 in). Gabriele says of her firing technique: 'I bisque fire over 24 hours to about 1000°C (1832°F) in a ceramic fibre gas kiln. My secondary smoke firing is done in a simple metal container (oil drum, dustbin, bucket, etc., depending on size of pot, one fired at a time), using sawdust, which I set alight from the top so that it smoulders down slowly, covering the container with a lid slightly ajar so that there is some airflow. In case of a large piece this process can last several days including cooling time. Depending on the sawdust I use and the way I pack the pots I can achieve fairly predictable results. I reckon the temperatures remain below 800°C (1472°F). While the bisque firing is oxidised, there will be some reduction and carbonisation in the smoke firing.' *Photo: Kelpie*

SMOKE FIRING

Smoke firing is very similar to pit firing although the top temperatures achieved are probably a little lower. This type of firing can be carried out in a brick pit or any metal or brick container that will allow enough air in for the fuel to burn. An old metal dustbin with a lid and several holes drilled into the base is ideal or a box constructed from house bricks.

To build a smoke-firing pit, place a layer of sawdust – or leaves, grass and suchlike – onto the pit floor, and place the first layer of biscuit-fired work on top. Add more sawdust and continue this layering of alternate work and sawdust until the bin is full with the last layer being sawdust.

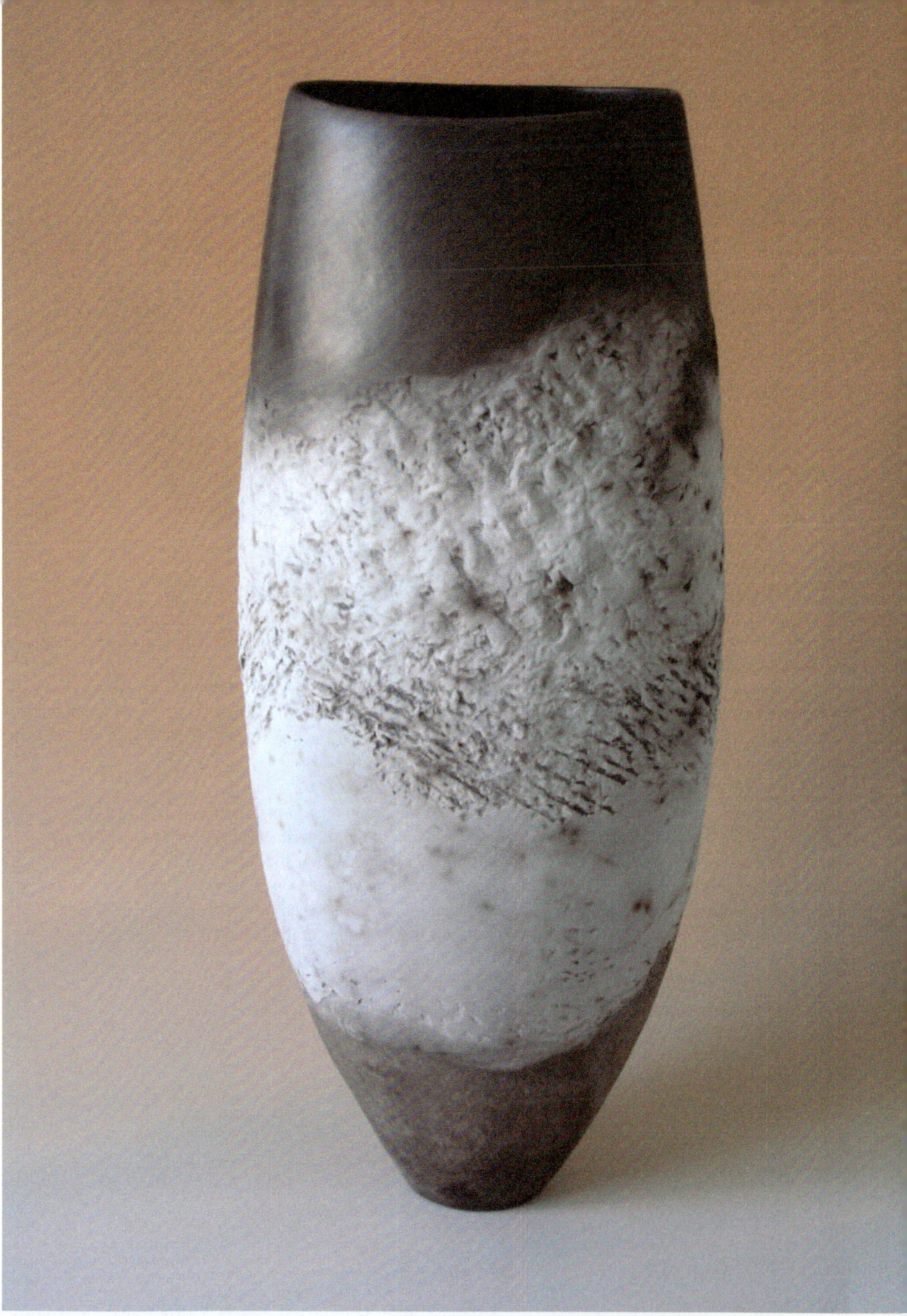

The grade of sawdust can be varied with finer sawdust tending to produce rich blacks, and coarser more open grades forming paler colours. Work that touches, can act as a form of resist decoration, as can the application of slip to the biscuit ware. Salt and metal oxides may be added to the sawdust mix as well as other combustibles which may contain active oxides. Banana skins are very high in potassium and can make some interesting marks as can dried seaweed. These can be scattered over the top or packed in amongst the work.

When the kiln is full, the top of the sawdust is lit and a cover loosely placed over the kiln. The sawdust should be smouldering not flaming as it burns. As the sawdust combusts, it produces a thin line of heat which probably reaches about 700°C (1292°F) locally. The biscuit-fired work will take on carbon, producing variegated surfaces. When the kiln is unpacked the work can be dusted down and then sealed with beeswax polish or similar, to bring out a warm sheen.

SAGGAR FIRING

A saggar is an umbrella term which describes a pre-fired ceramic flat-based vessel, usually constructed as a round or oval container made using a high-alumina fire clay or crank body. This container traditionally either wholly or partially protects work from undesirable fumes, gases and debris from the firing fuel. A saggar can, however, also be used to form localised atmospheres around particular pieces of work. This technique is best used in fuel-burning kilns, although if the saggar is well sealed this could be carried out in electric kilns.

The piece of work is placed inside the saggar with wadding lifting the piece off the floor, and then packed with combustible or reactive materials such as sawdust, wood ash, wire wool, banana skins or colouring oxides. A soft clay coil can then be applied as a seal and a kiln shelf or lid placed over the top. The kiln is then warmed gently to dry the coil and fired up to the desired temperature. If sawdust is used and the kiln fired to high, stoneware temperatures, then ash from the wood may form an ash glaze; this can be promoted by the addition of a small amount of borax in the sawdust mix. Wire wool will also melt and combine with the clay surface. Within the sealed saggar the combustion of materials placed inside can generate a localised very heavy reduction atmosphere and this propagates the fluxing of iron in the clay body, as well as reducing any glaze which may contain metal oxides that respond to a reduction atmosphere (such as iron and copper).

If the kiln is fired to lower temperatures, work may be carbonised by smoke developed by the burning of combustibles. A terra sigillata or burnished surface may show this off to good effect.

RIGHT: *Perforated Saggar,* Benedict Brierley. Thrown from industrial crank clay and perforated to allow a small amount of carbon and flame borne alkali onto the vessel inside, used in anagama wood firing. *Photo: Author*

BELOW: *Vessel with quilled lid,* Christine Gittins, height: 22 cm (8½ in). Christine's pots are thrown, burnished and biscuit fired to 1000°C (1832°F). They are then packed into saggars with salt, copper and sawdust and fired in a gas-fired raku kiln for 6 hours to 850°C (1562°F). *Photo: Christine Gittins*

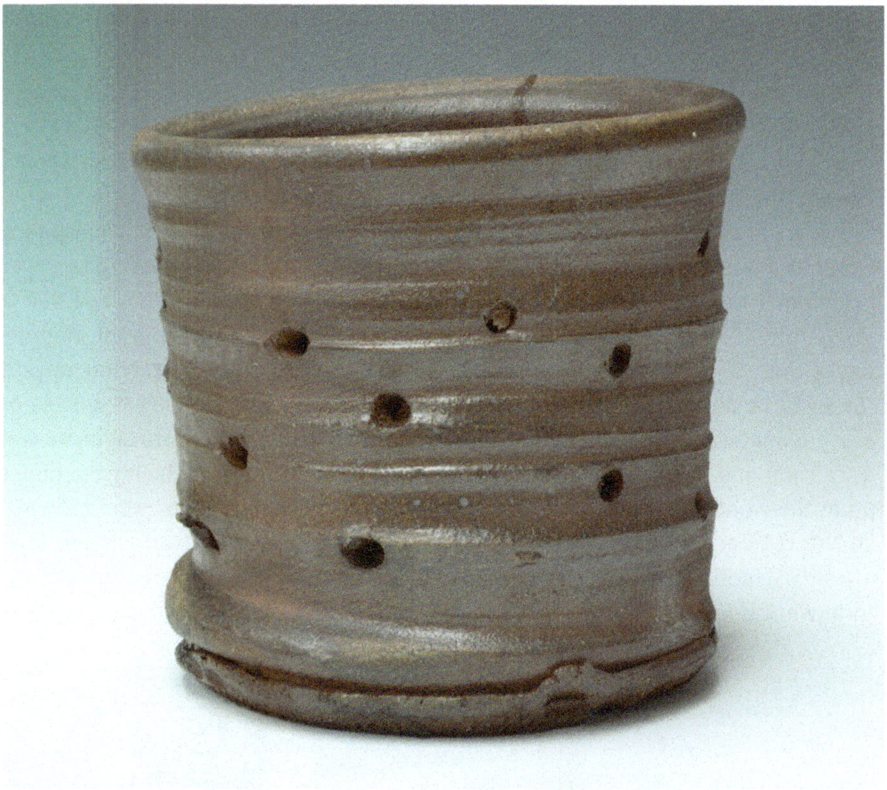

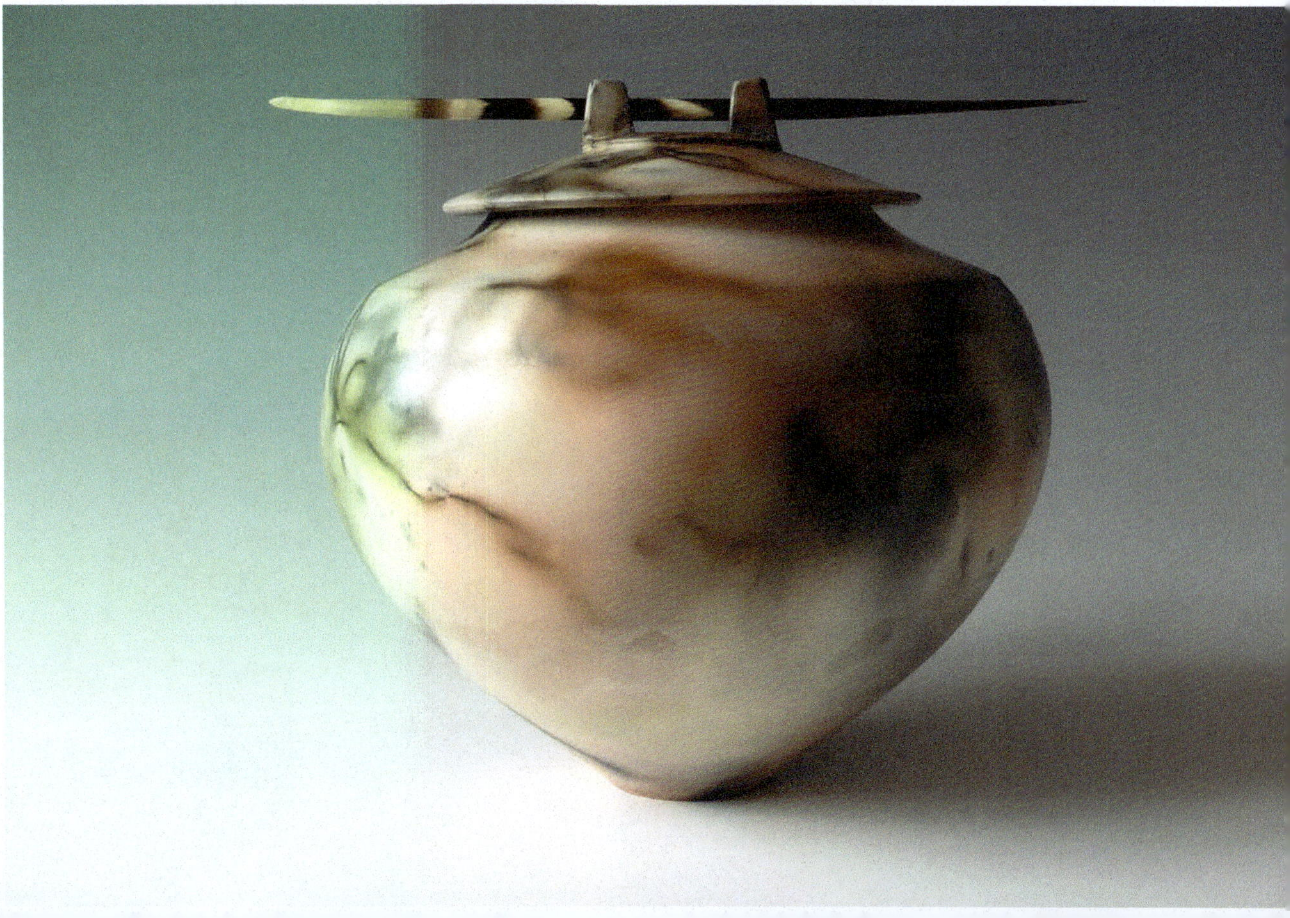

WOOD FIRING

Firing kilns with wood as the main fuel may also be placed into the specialist category of firings. For, as well as being a fuel, it also produces bi-products from its combustion that are capable of interacting with raw clay surfaces and applied glazes, and modifying them. The degree to which the decorative effects of wood firing are visible depends largely on the length of the firing, the atmospheres generated and the types of clays being fired in the kiln.

In wood firings as the wood combusts, ash and alkalis are liberated, and due to the nature of wood kilns as described in earlier chapters the draft from the chimney draws these through the kiln. When the clays reach low stoneware temperatures, the surfaces start to become tacky so the ash will stick as well as being deposited on any horizontal surfaces. The soluble alkalis, such as sodium and potassium, as well as calcium and metal salts, such as iron and magnesium – all of which were bound up in the wood (drawn up from the earth in which the source trees were grown) – also start to accumulate and modify the clay surface. If a significant enough layer of ash is deposited on the work over the duration of a long firing then the ash will form what is known as a natural fly ash glaze and begin to run, forming rivulets of

BELOW LEFT: 'Smokeless' wood-fired kiln, Sheffield, UK. A small fire is set in the firebox chambe beneath the grate (left-hand side of the kiln). When the firebox is hot enough, wood is fed onto a grate below the stok hole on the left at the front of the kiln. Heat is then drawn into the packing chamber (on the right of the structure), and pulled up through the work (Bourry box-style firebox). Fumes escape through the tall chimney. *Photo: Author*

BELOW RIGHT: Early stages of wood firing in a crossdraft kiln Matthew Blakely's kiln being warmed up with a small fire outside the lower stoke hole. *Photo: Matthew Blakely*

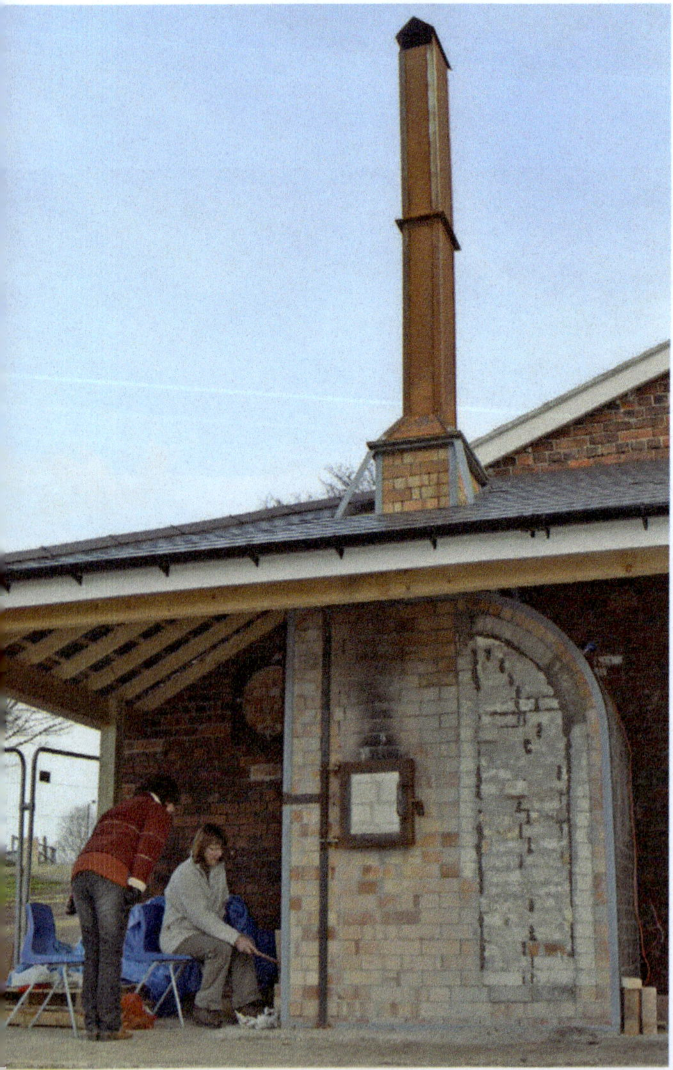

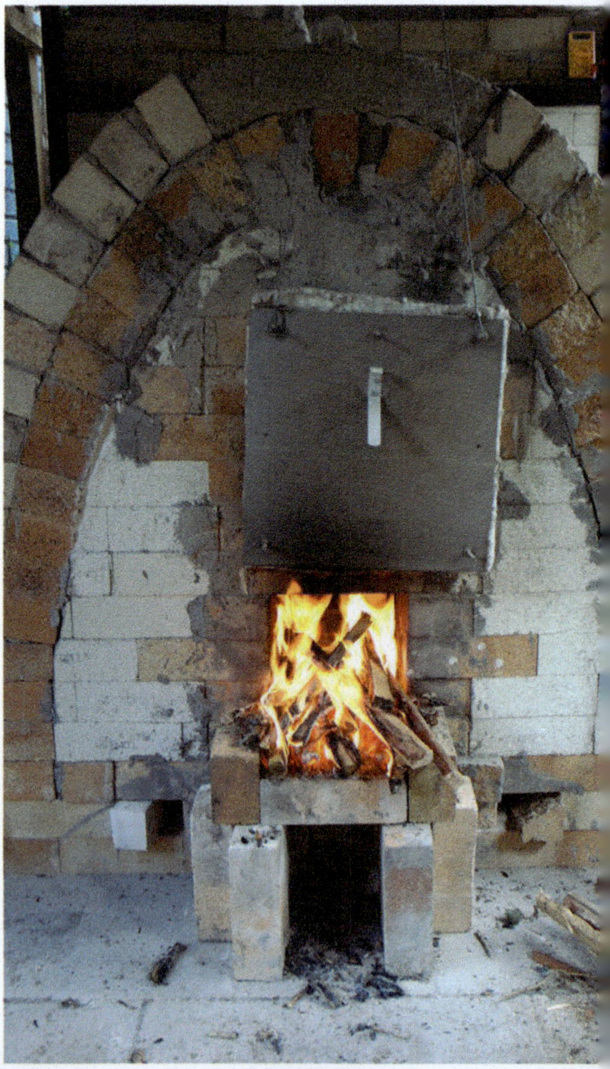

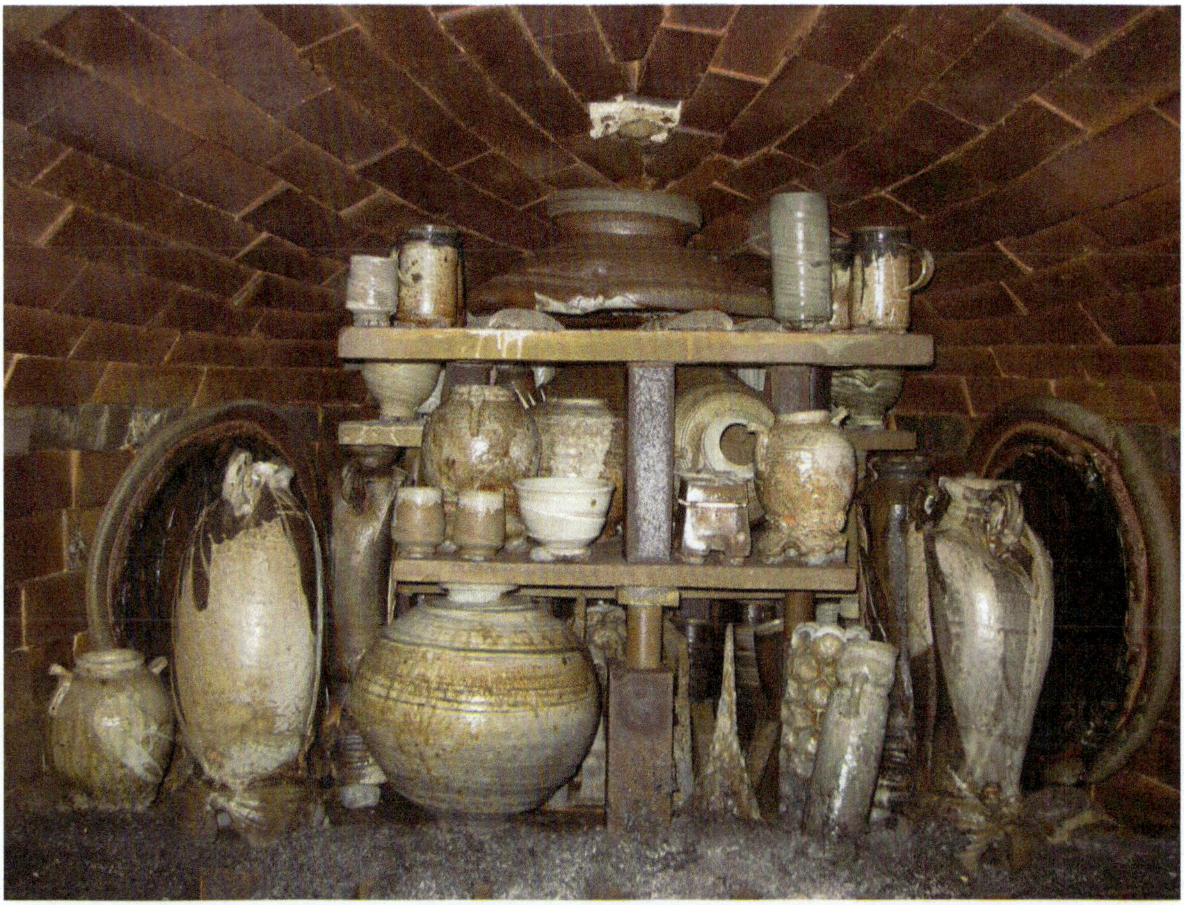

ABOVE: Unpacking the back end of the Loughborough anagama kiln after a 4-day firing. *Photo: Author*

glaze. Natural fly ash glazes will tend to be greeny blue in colour, depending on the minerals contained within the ash.

Surfaces not directly in the flame path will be influenced by the passage of alkalis, but will also be influenced by the atmospheres generated. Within the chamber of a wood kiln the flame is moving directly through the work so the micro atmospheres generated can vary with quite small distances to the extent that two pieces standing next to each other on a shelf and made from the same clay may come out from the firing looking radically different. It is worth noting that work going into this type of firing should be placed on wadding. This varies in composition slightly from the wadding used in salt and soda firings but performs the same job.

Wood-firing wadding

2 x sand
2 x fire clay
2 x china clay
1 x sawdust

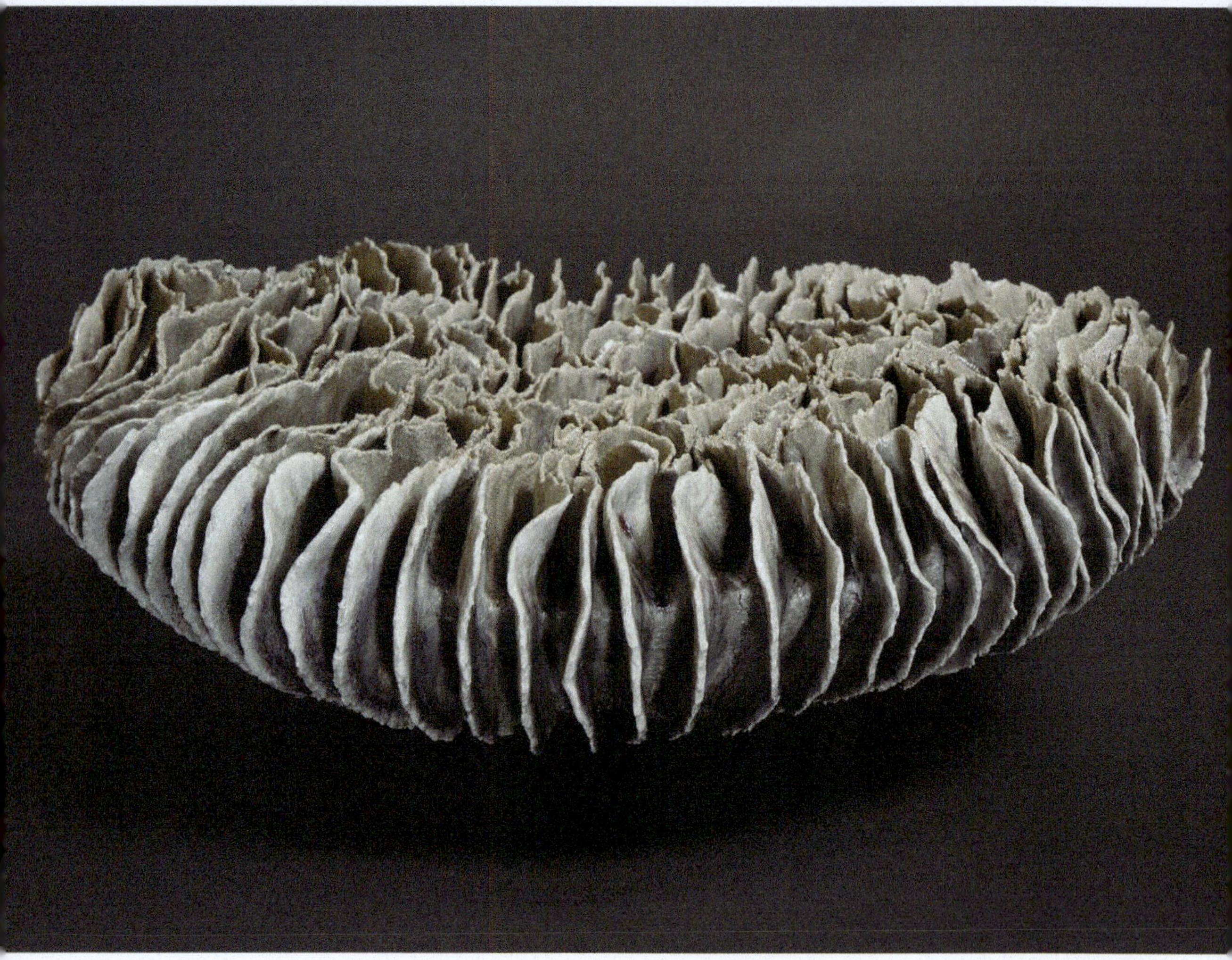

ABOVE: *Gilled Bowl Form*, Sarah Villeneau. Fired in the Sheffield smokeless wood-fired kiln to cone 10 (1280°C/2336°F), 33 x 10 cm (13 x 4 in). *Photo: James Mulkeen*

RIGHT: *Dartmoor Blues Vessel*, Matthew Blakely. Height: 38 cm (15 in). Wood fired to 1280°C (2336°F) in his own kiln, in reduction. *Photo: Matthew Blakely*

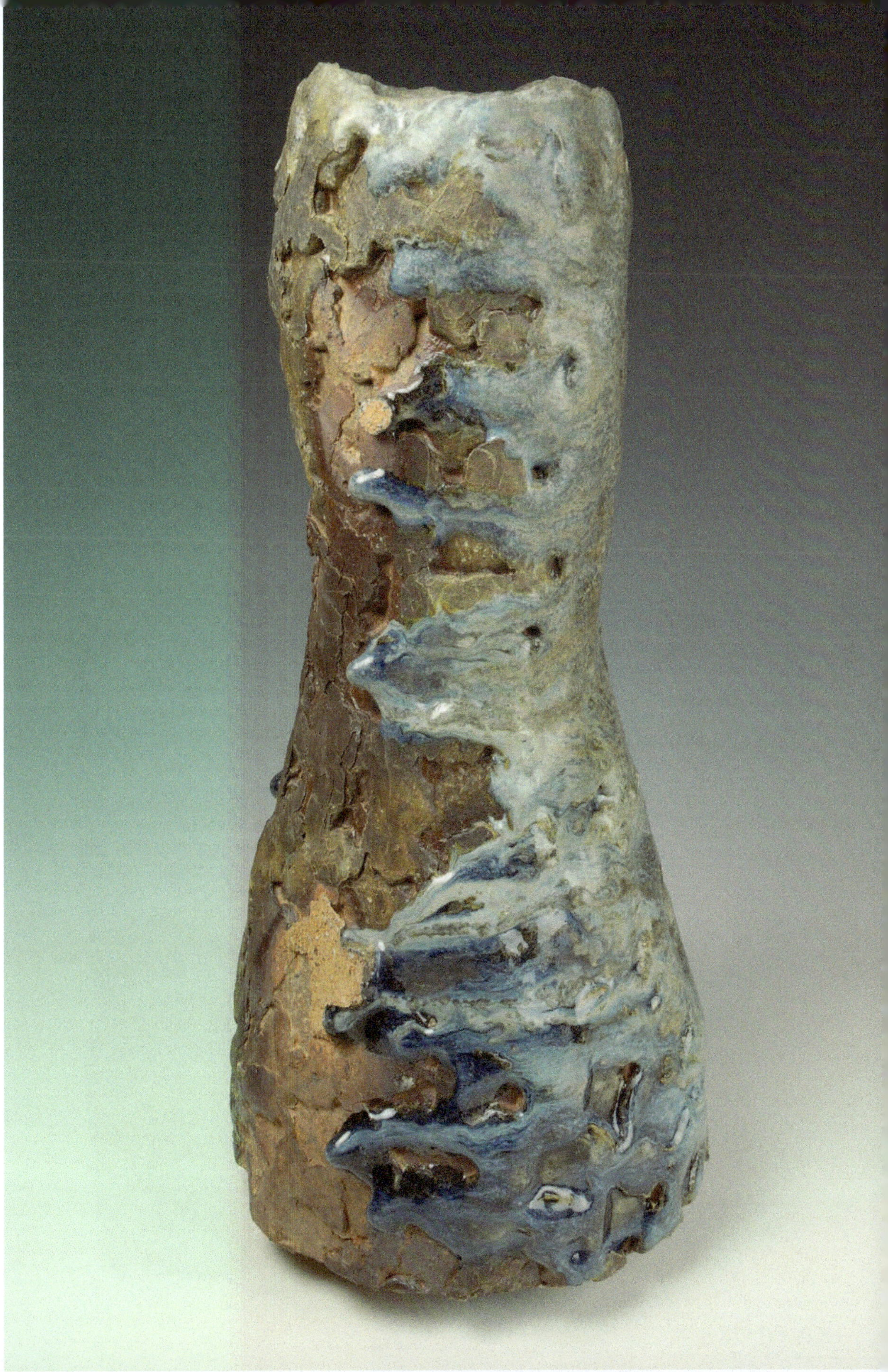

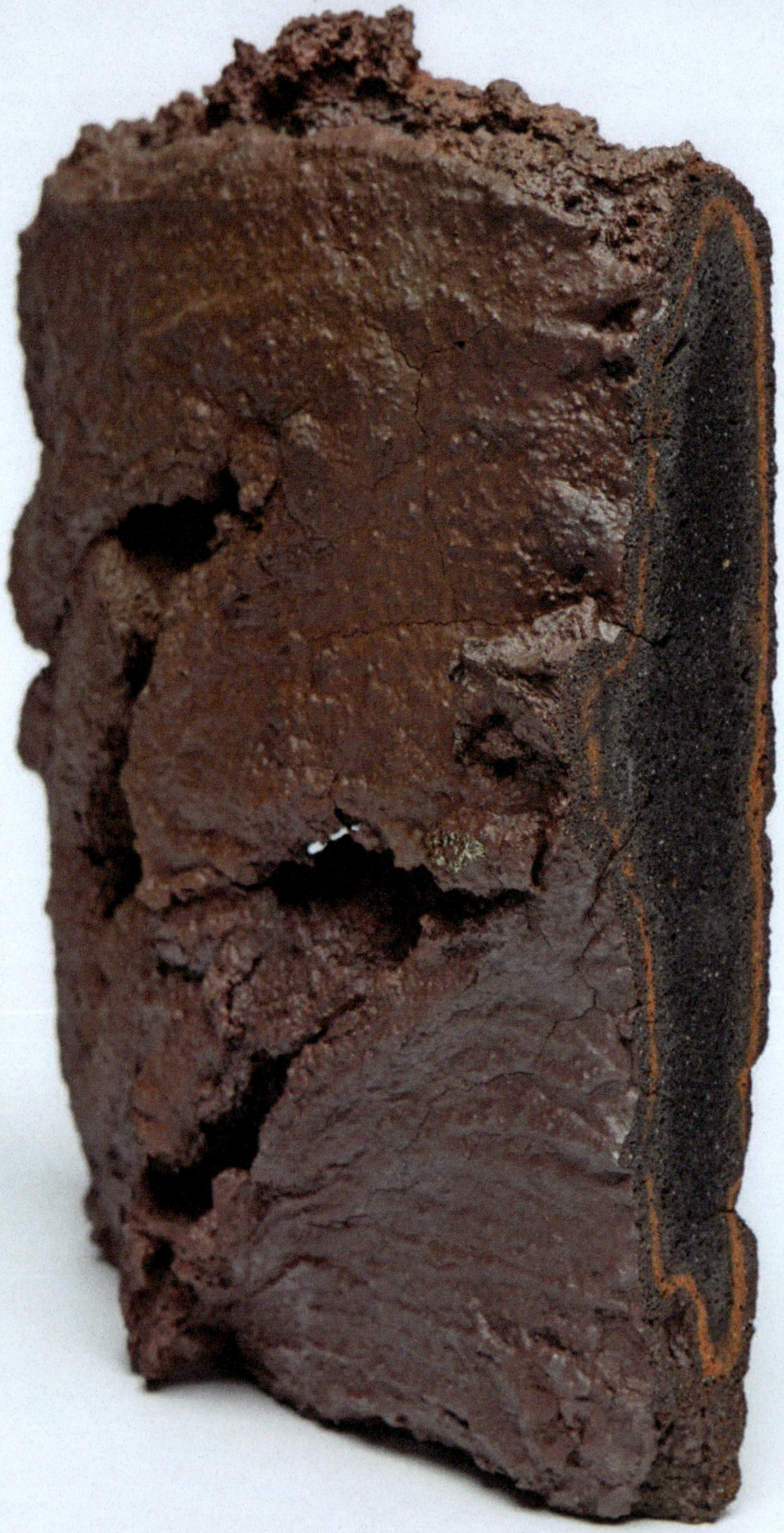

6 Firing faults and remedies

Clay can be a fickle partner. No matter how well conceived the form, or how meticulous the making, all can be undone in the firing. Even if a piece survives the initial biscuit, there can be potential problems which only become apparent in the final glaze firing, turning the anticipation and excitement of a kiln opening into a damp squib of disappointment. This chapter looks at some of the potential problems encountered when firing and suggests some ways in which these problems may be avoided.

AIR BUBBLES: THE MYTH

It is a common misconception that air pockets in clay will cause it to explode in the biscuit firing; this is a bit misleading. On its own, an air pocket will not cause a piece to explode. The principle cause of damage to work in the early stages of biscuit firing, assuming that the clay is properly prepared and does not contain any contaminants, is residual moisture; in other words, the work has not been sufficiently dried prior to the firing. What moisture will do in the drying process is migrate towards voids in the clay before finally being released to the atmosphere. Therefore, an air bubble may become a pocket of moisture and the epicentre of a fracture.

The degree to which something will explode is relative to the amount of water present. If the moisture is localised then it may only be a small piece of the surface which becomes detached; if, however, the whole piece is damp, then the result could be a pile of rubble. The mechanics of why this happens is straightforward: when water is heated above its boiling point it turns to steam and expands massively, taking anything which encases it with it. So avoiding air bubbles is a simple solution to minimise the risks.

WATER

There are several ways of ensuring that work is not lost in this early stage of firing. A good way of testing whether a piece is dry is to gently heat the piece in question for a short time on a radiator or airing cupboard and then hold against a cold exterior window; any water present will be seen as condensation on the glass. If this happens then dry the work for longer. Even though work may look dry on the surface there may still be moisture present deeper into the body. This is particularly relevant to dense un-grogged bodies such as smooth terracotta, porcelain and some stoneware clays. Grogged bodies will tend to dry more rapidly as the water is able to escape through the more open clay fabric.

LEFT: *Ceramic Object,* Jim Gladwin. Gas fired to 1200°C (2192°F) in reduction, 30 cm (11¾ in). This piece has been cut with a brick saw after firing to juxtapose the internal clay fabric with the external clay surface. *Photo: Alan Duncan*

An additional precaution to avoid drying problems is to fire the kiln gently in the early stages of a biscuit firing. It is generally accepted that a climb of 100°C/hour (212°F/hour) is appropriate for a biscuit firing. If it is suspected that the work may be a little damp, or if the work to be fired is thick or made of a dense un-grogged clay, the firing may be taken even slower at 50°C/hour (122°F/hour) for example until the kiln is at 600°C (1112°F) and then 100°C/hour (212°F/hour) to the desired biscuit temperature. The kiln may also be soaked at around 90°C (194°F) for several hours so that any moisture can evaporate before the kiln begins to climb.

This is a simple procedure in an electric kiln with a programmer as long as the programmer attached to the kiln is of a multiple ramp type. If, however, the controller has only two ramps it may be necessary to just do a low-drying ramp, and then reset the kiln to do the biscuit firing proper. If a manual electric kiln is being used, setting the energy regulator to 5–6% should be enough not to allow the temperature to climb too high, but vigilance will be the key to ensure that the kiln stays below 100°C (212°F) until the work is dry. If in doubt, remove and carry out the window test. Alternatively, holding a cold piece of glazed tile over the kiln's top vent will produce condensation if water is still being driven off. A gas kiln should be treated in a similar way; igniting the burner pilot light, or lights, should provide enough warmth to dry out work before the firing is started proper. However, vigilance is the key, keeping an eye on the kiln's pyrometer to ensure the temperature remains below 100°C (212°F), or using the glazed tile test.

It is also worth noting that even when a piece of work has been dried thoroughly, if it is then subjected to a damp environment it will reabsorb some atmospheric moisture. This could become a problem if one's kiln is situated outside or in a damp shed and the door of the kiln is left open overnight. Work in the kiln may absorb some dampness and so need to be warmed gently and fired appropriately. This may also apply to biscuit

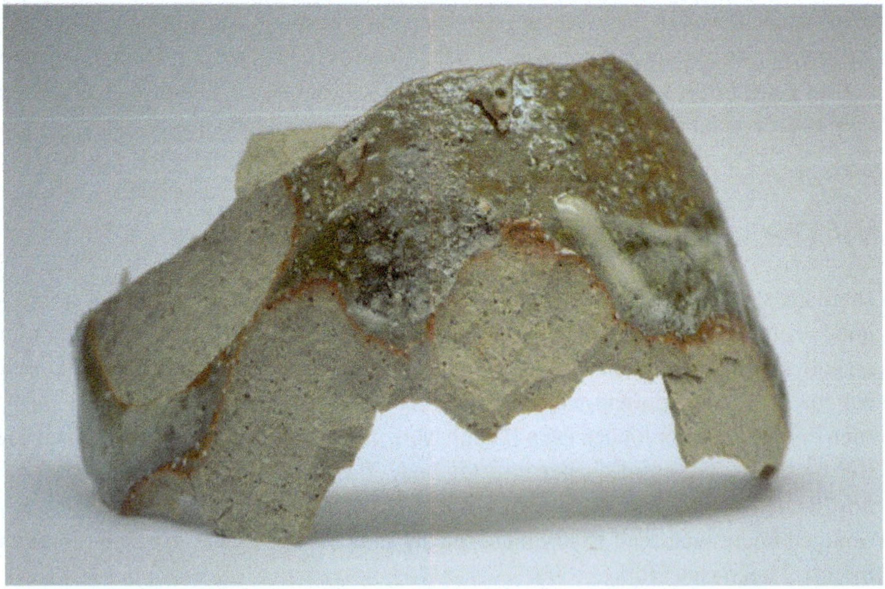

Ceramic piece damaged due to water absorption, leaving a characteristic rough, slightly dimpled surface. In this example, the piece had been biscuit fired, glazed and then heated too quickly. Water absorbed from the glaze caused the piece to explode. *Photo: Author*

ware which has been left in a damp environment before firing; again gentle warming is the best way forward. Care should also be taken to ensure that work is allowed to dry after a glaze has been applied; as some of the glaze's water content is drawn into the bisque body, this water must be allowed to escape before the kiln is allowed to exceed 100°C (212°F).

DUNTING

'Dunting' is a term which describes a particular type of cracking of the ceramic body. Cracks may be formed in biscuit firing if the work is damp, or in the final glaze firing due to trapped gases; they may also be unwittingly created in the clay body at leatherhard stage if too much pressure is applied to a piece. With this type of cracking, the piece may emerge from the biscuit firing seemingly fine, but the crack will only show itself in the finished piece. The crack may not be visible, but the fault can be identified by gently tapping the piece; if it does not 'ring' but sounds dead, dunting is likely to have occurred.

Dunting tends to be caused by firing or cooling too fast. It may happen either as the kiln temperature climbs or cools, and is more common in high-silica bodies or smooth bodies which contain little or no grog. Other contributing factors may be: uneven clay wall thickness, poor glaze fit, uneven glazing overall but specifically between the interior and exterior surfaces, and uneven heating. All of these can cause stresses in the clay body making them vulnerable to dunting.

Dunting tends to occur at two specific points: first, between 570°C to 580°C (1058–1076°F) as the kiln climbs and cools; and second, around 220°C (428°F) as the kiln cools. As temperature increases, there is a large expansion in the silica crystals contained within the clay equating to a linear expansion of roughly 0.45% as silica converts from

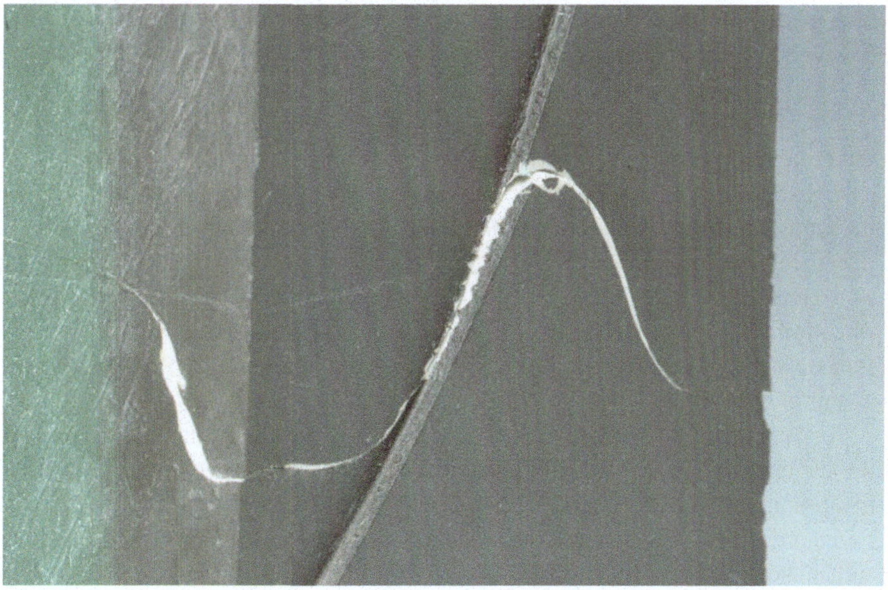

Body dunt in a slip cast semi porcelain clay. *Photo: Author*

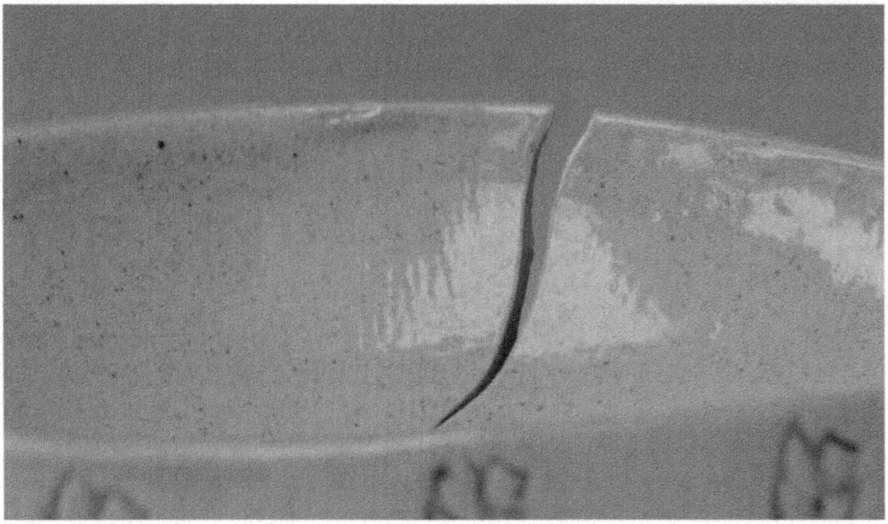

Rim dunt in a press-moulded bowl made from a fine white stoneware clay. *Photo: Author*

alpha to beta quartz. This reaction is reversible and a similar rapid contraction occurs when the kiln is cooling. On cooling, cristobalite is also inverted, which causes a quick contraction; this can help with the compression of the glaze, but can also stress the ceramic to the point that it fractures.

These sudden expansions or contractions can shock the clay body and cause cracking; from hairline cracks to substantial cracks, which distort the clay on either side. It is possible for a hairline crack to be formed in the biscuit ware and go unnoticed until the final firing where it may open up to become much larger. This is most noticeable when the final firing is to stoneware temperature due to increased contraction of the body. On a glazed piece of work, it is possible to tell if dunting has occurred on the way up or the way down by examining the edges of the crack closely. On a piece of work which has dunted as the kiln was climbing, the edges of the crack will tend to be more rounded as the glaze has softened and moved slightly to cover the sharp ceramic edge. If the edges of the crack are sharp, then the dunting is more likely to have occurred as the kiln was cooling; the glaze has already solidified and therefore has a sharp edge when broken.

It is possible to hear if a piece of work has a crack in it by tapping or flicking it in the same way as one might check a kiln shelf before packing. If the piece rings then all is well, if a dull burring sound is heard then the piece is almost certainly flawed. It is worth mentioning that a crack can also be inadvertently formed if the clay is stressed when in the leatherhard state. If the clay is manipulated too extremely while at this stage, a fault may be formed in the body which may go unnoticed until either the biscuit or glaze firing where it will pull apart to form a crack.

To avoid dunting, the most important aspect of firing is to be mindful of allowing the kiln to pass gently through the temperature ranges where the ceramic is susceptible to dunting: roughly 200°C to 300°C (392–572°F) and 500°C to 600°C (932–1112°F). As stated previously, a temperature climb of 100°C (212°F) should be sufficiently cautious and applies whether the kiln is fuel fired or electric fired. It may also help to position

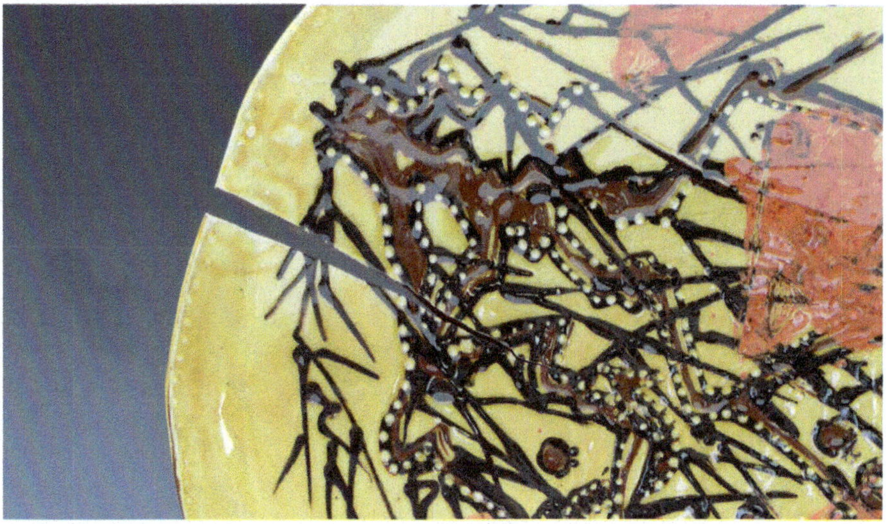

Glaze-firing crack. The crack was formed by stress in either the leatherhard or dry body. *Photo: Author*

susceptible pieces – those with wide bases or taller pieces – in the centre of the kiln so that heating and cooling are as even as possible; this will at least diminish stress on the work caused by uneven heating and cooling.

BLOATING

Blistering or 'bloating' can be caused when the gases emitted from the clay as it is being fired are unable to escape effectively through the clay body wall. This fault is fairly uncommon, especially as so many of the clay bodies which are used in studio ceramics today are very processed, which removes many of the impurities that produce gases in decomposition. However, there are a number of scenarios which may contribute to the likelihood of it occurring.

First, if the clay is fired too close to its vitrification point too fast, the surface of the ceramic may become too fused for the gases to escape, yet, due to pyroplasticity, are soft enough to distort under the pressure exerted by the build-up of gas under the surface; thereby causing the surface to blister. Second, if the biscuit firing is undertaken too rapidly, or the biscuit firing is not high enough, unburnt carbon can be left in the biscuit ware. When the subsequent high firing takes place, the carbon cannot escape and causes the body to bloat. Third, if work has been glazed all over and the biscuit firing was too fast, the glaze layer can also trap gases within the body; this is more common with earthenware bodies and is one of the reasons that many old red earthenware vessels were only glazed on one surface so that any gasses were able to escape from the unglazed side.

The gases that are mainly responsible for bloating are carbon dioxide, carbon monoxide and sulphur. Carbon dioxide and monoxide start to be given off between 700°C to 900°C (1292–1652°F) and sulphur in the 1100s°C (2000s°F). It can clearly be seen then that if a biscuit firing is too low, there is a strong chance that unburnt

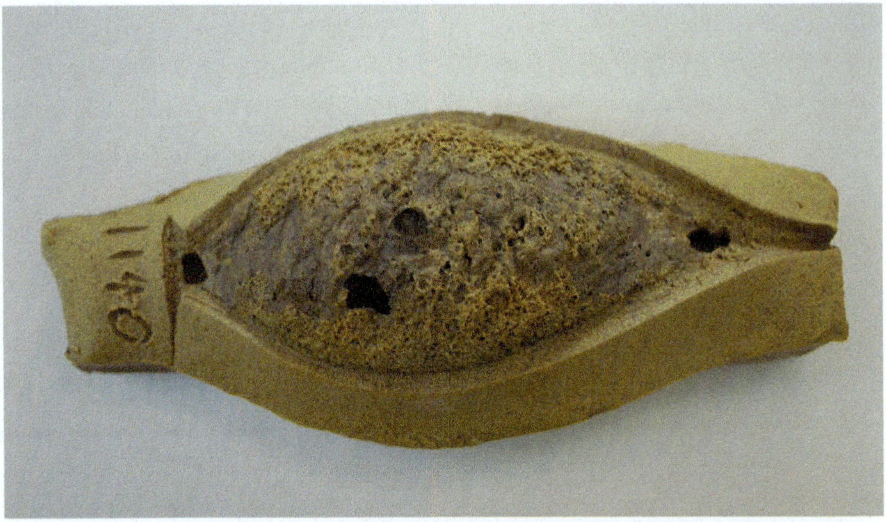

Bloating in a low-firing earthenware body. This locally dug clay bloated at 1140°C (2084°F). *Photo: Author*

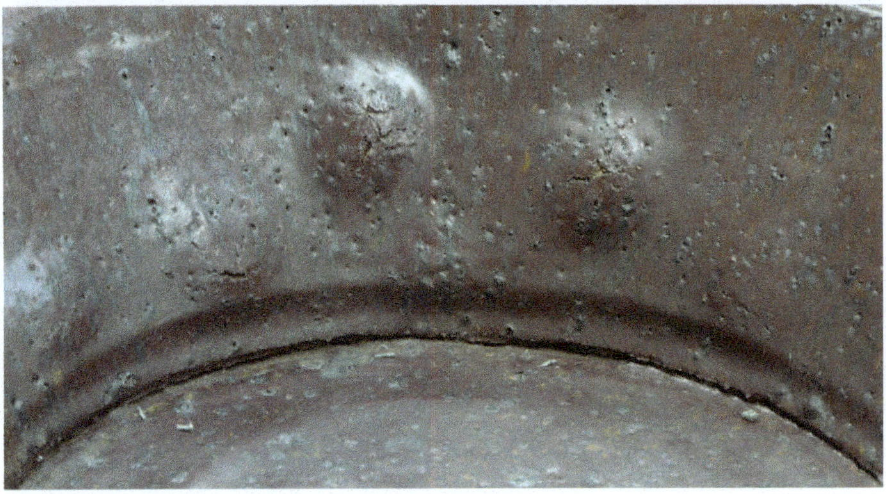

Impurities in this brick clay bowl have caused it to bloat. Fired in the front of an anagama kiln, the surface fused and trapped gases in the body. *Photo: Author*

carbon may be left in the body. If the subsequent glaze firing was then to be carried out too quickly, this carbon may not be able to escape, which can cause flaws. The problem of bloating may be exacerbated in a reduction firing. The mechanics of this are that carbon in the body loses oxygen which bonds with iron in the clay at stoneware temperatures; as stated previously, iron becomes a strong flux and it is this strong fluxing action that can seal the external surface of the clay, thereby trapping gases.

Firing cautiously is the best way of avoiding this type of problem. As with dunting, best practice is to take biscuit firing gently especially in the early stages and ensure that the firing is high enough; generally from 980°C to 1000°C (1796–1832°F) is a good low-biscuit temperature. If intending to apply an earthenware glaze, then a high-biscuit firing from 1120°C to 1180°C (2048–2156°F) will be used anyway, so the problem of too low a biscuit firing shouldn't be a problem. Due to the release of

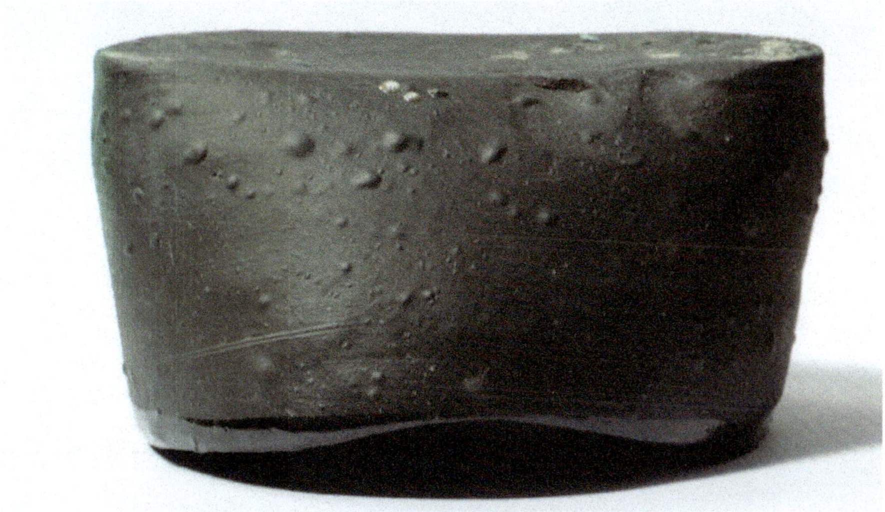

Black body: starting to blister due to over firing. Fired to 1260°C (2300°F) in oxidation. *Photo: Author*

sulphur at higher temperatures it is prudent not to rush this part of the firing either; especially when carrying out a reduction firing, as iron in the body will become an even stronger flux, which can exacerbate the potential problem of the surface becoming vitrified quite early in the firing.

In some of the clays that the author uses in protracted anagama firings there is an unusually high flux content in the form of nepheline syenite, which is a soda flux that melts earlier than feldspars. This inclusion encourages the flame borne alkalis from the wood fuel to form a bright fly ash glaze surface. Pieces almost self-glaze themselves in areas that are devoid of applied glaze. These clays can be seen to have sheen in the mid-1100s°C (2100s°F) temperature range. It is important, therefore, that these clays are heated slowly, as the surface fuses relatively early in the firing, which ultimately can be as high as the upper 1300s°C (2500s°F).

If any clay is fired above its vitrification point, the clay will start to break down and eventually melt.

BLACK CORE

'Black coring' is a fault which again can result from the outer clay surface vitrifying earlier than the centre. The problem usually, but not exclusively occurs in reduction firings or in biscuit firings carried out in fuel-burning kilns which are allowed to reduce. The fault generally occurs due to lack of oxygen between the temperatures of 700°C to 900°C (1292–1652°F), and again is a result of the firing being carried out too fast. If carbon is not able to oxidise to CO_2 it liberates oxygen from any iron present in the clay. The Fe_2O_2 form of iron is reduced to FeO which is a powerful flux. The fluxing action of FeO can seal the clay surface, thereby trapping the unburnt carbon which is then visible as a black core in the centre of the ceramic wall cross section.

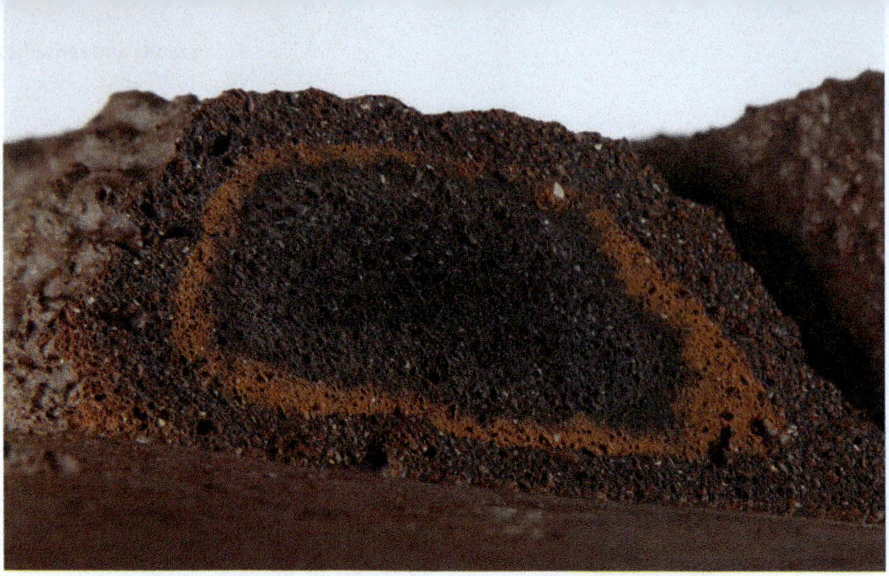

Cross-section of fired ceramic displaying black core. The orange line is the original clay colour with the black core on the inside and reduced ceramic surface on the outside. *Photo: Alan Duncan*

PYROPLASTICITY

'Pyroplasticity' is a phenomenon which occurs at the upper end of clay's firing range. At these temperatures the ceramic softens, almost as soft as when it was in a plastic clay state, which can make some forms vulnerable to collapse due to gravity.

Weak forms and clay bodies which have higher flux levels are usually the most susceptible to the effects of pyroplasticity. For example, a wide bowl – especially if made from a highly vitrified body such as porcelain or bone china – could at the higher end of its firing range slump or distort. In this situation the problems can be overcome by careful planning. If the form is structurally weak then it may be necessary to construct a setter to support the object in the firing. A setter is simply a supporting structure in the kiln; ideally a bespoke setter is made from the same material as the object it is supporting, thereby addressing any issues which may develop through differing shrinkage rates.

When finely cast pieces are fired such as a bowl or similar, it is sometimes necessary to create a setter to avoid any distortion in the rim. In this instance a conical setter could be cast, and the bowl placed on it upside down. The setter is conical so that the object being fired can slide up on the cone as shrinkage occurs. Some ceramic artists embrace this pyroplastic characteristic to introduce a little uncontrived movement into a piece of work by intentionally altering their control of the firing. Many faults can be likened to weeds; that is, a weed is simply a plant which is growing in the wrong place. What some makers may view as a fault others may see as a desirable quality just beyond the control of the maker.

DISTORTION/WARPING

Distortion and warping in the kiln can be attributable to pyroplastic movement in higher-temperature stoneware firings. However, it may also be caused by several other factors.

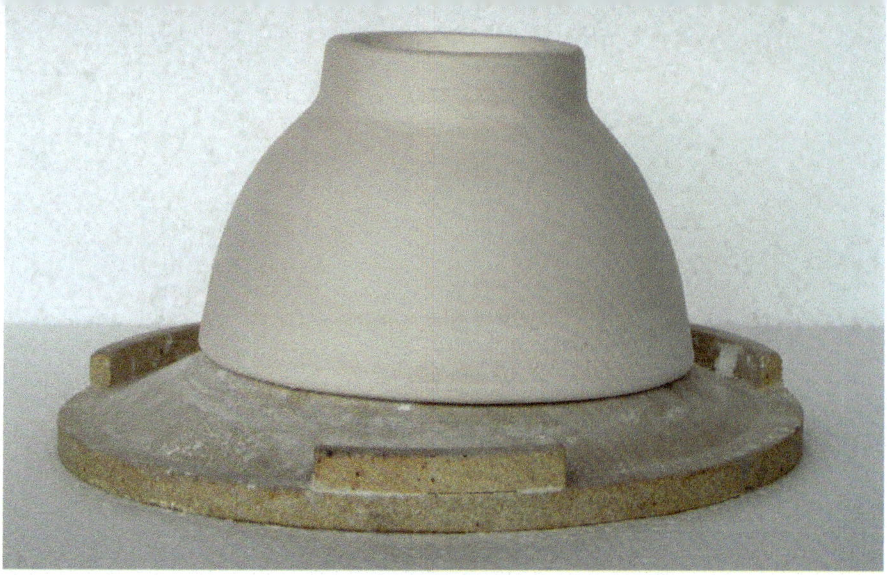

Small bowl placed on a kiln setter. Used for reducing rim distortion when high-biscuit firing. *Photo: Author*

Clay memory

Clay has a making memory which can manifest itself in the firing. An example of this could be the way seams on a fettled piece of cast work may be invisible when the piece is dry, but become apparent only after the first firing. This is caused by stresses in the material when it is formed, relaxing in the firing and echoing an original shape. Although not really a firing fault, allowing a high firing to dwell in the mid-1100°C (2100°F) region of a firing may propagate the problem. Warping and distortion may also be caused by uneven heating of ceramic in the kiln, as further shrinkage takes place during the firing. If this happens unevenly, it can cause the piece of work to distort. In extreme circumstances the piece may crack.

GLAZE PROBLEMS ASSOCIATED WITH FIRING

Although not strictly firing problems, some glaze faults can provide evidence of the way that a kiln should be fired for a specific type of firing. As an obvious example: if a glaze that is known to be fluid and glassy comes out from an electric kiln firing with a matt finish, clearly something has gone amiss. The matt finish would most likely indicate that the kiln has not reached the required glaze melt temperature (assuming of course that incorrect glaze mixing has been ruled out), in which case it should be investigated to ensure that there is not a problem with any of the elements in the kiln. Investigating the firing could be undertaken by placing the appropriate cones in the kiln so that there is a visible record of the actual temperatures reached after the kiln has cooled. Likewise, if cones are not used, in a fuel-burning kiln the thermocouple and pyrometer should be checked that they are correctly calibrated and giving an accurate reading.

Glazes coming out of a firing with soft-edged craters or bubbles would also usually indicate that the kiln has not reached temperature. If, however, the bubbles have sharp

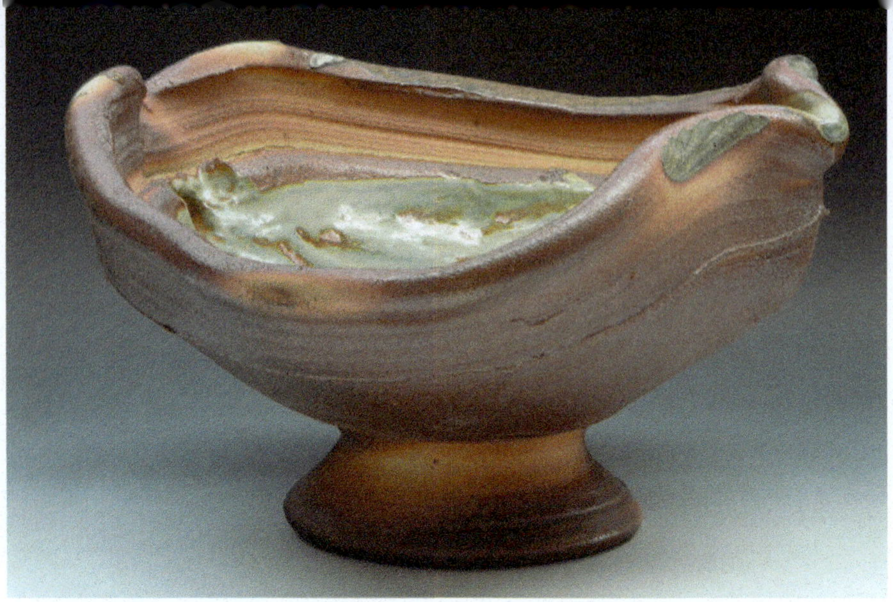

Large Side-fired bowl, Benedict Brierley. Fired in an anagama wood-fired kiln for 3 days to 1400°C (2552°F), diameter: 46 cm (18 in). Pyroplasticity has been used to distort this bowl adding to the sense of movement. By firing on its side, the weight of the bowl has distorted the edge which was wadded on the floor of the kiln. This also allowed the ash glaze to flow with gravity. *Photo: Author*

edges this could suggest that the glaze has boiled and that the kiln has over fired. The remedy would again be to investigate the above and ensure that the kiln is firing to the required temperature.

CAUSES OF BROKEN KILN ELEMENTS

As has been stated earlier, kiln element wire can vary in thickness depending on the size of the kiln which is to be heated. The larger the kiln then the heavier duty the elements tend to be. However, irrespective of the gauge of wire used, kiln elements do fail. There may be several causes:

1. If the kiln is not cleaned out regularly, pieces of ceramic may come into contact with the kiln element. Ceramic contact with element wire can cause a hot spot to form, which will weaken and eventually cause the element to break; this will be evident by a small black mark and slight melting around the mark on the element.

2. If the recessed channels that the elements are housed in are not well maintained. While elements are being heated they become soft, so if they are not supported by the brickwork they can slump and start to hang down. After several firings, they may stretch to the point that they simply snap or they may come into contact with the element below and short out the element circuit.

3. If glaze is allowed to touch elements, for example if a fluid glaze runs and fills floor element grooves. Assuming the glaze has not got into the elements and that the element has broken as a result of a hot spot, the element may be temporarily repaired.

How to temporarily repair a broken kiln element

In order for the element to be manipulated, it must be heated locally. The element will become soft and stretchy when at orange heat; this temperature may be reached using a small handheld brazing torch or similar. Elements should not be handled roughly at room temperature, as once they have been fired they become brittle, and attempts to stretch them will cause them to snap. Once the local area around the element break has been heated to orange heat the two ends of the element can be straightened with pliers. Once straight they can be twisted around each other to reform the connection. If the twisting is as tight as possible this repair may last for many firings; at the very least, it should keep the kiln running until new replacement elements arrive.

As alluded to earlier, the brick work within the kiln chamber and surrounding the elements should be kept in good condition. HTI brick which most modern electric kilns are made of is not very hardy, and may be quite easily damaged if banged by work or kiln shelves while placing pieces into the kiln. Small dents and chips will not affect the running of the kiln, but if element channel walls are damaged, then they should be repaired. Having a few spare HTI bricks to hand and some refractory cement will make repairs fairly painless. With the kiln switched off, the old brick can be cleaned out and a new piece cut from the spare brick; this may then be cemented into place. Once fired, the mend will be as good as new.

GAS KILN FAILING TO ACHIEVE TEMPERATURE

If the temperature fails to climb when carrying out a gas firing, there are several things that can be checked to attempt to resolve the problem.

1. Check that the burners are not blocked. These should be checked before each firing anyway, especially if they are of the type entering the kiln through the floor. It is not uncommon for bits of ceramic from a blowout to fall into the burner pots and restrict the gas flow.

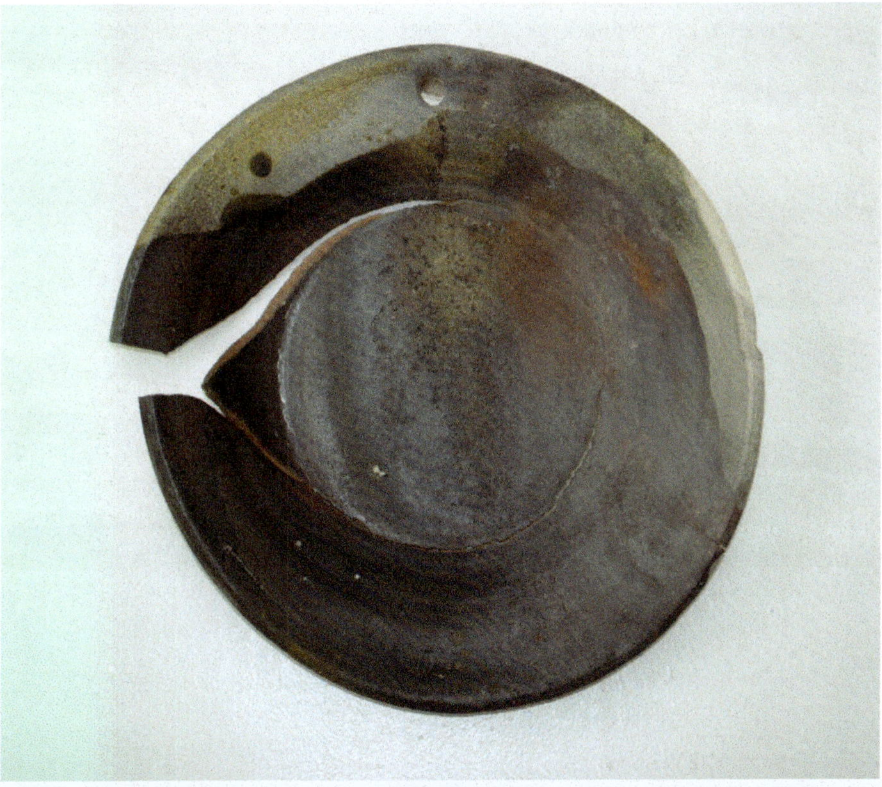

Warped platter: the dunting and distortion of this piece was caused by significant temperature difference from one side of the platter to the other. The platter was fired in the firebox of an anagama wood-fired kiln, and half buried by embers for most of the duration of the firing. This effectively insulated one half from the ambient temperature of the rest of the kiln. The temperature differences involved were probably between 900°C (1652°F) on the buried half, and 1320°C (2408°F) on the upper half. *Photo: Author*

2. Ensure that there is sufficient gas pressure to complete the firing. This is more critical when using propane gas from cylinders. As the level of gas in the cylinder drops, so does the pressure and there may not be enough pressure to generate the amount of heat needed to finish a firing. If this is the case, a spare full bottle should be on hand so that the firing can be completed with a fully pressurised bottle.

3. Before and during packing, make sure that there is no restriction to the flue, both in the chimney and (in the case of a downdraft kiln) in the flue which runs under the base shelves. If exhaust gases are not able to fully escape from the chamber the fuel will not burn efficiently and not enough heat will be generated.

4. Ensure that there is not too much fuel entering the kiln or that all air inlets are clear. Either of these can cause the kiln to stall as the fuel will not combust efficiently enough to produce enough energy for the kiln to climb in temperature.

WOOD KILN FAILING TO REACH TEMPERATURE

Unique to this type of kiln, the fuel itself may be a cause for poor or lack of temperature climb. Differing species of tree will produce varied amounts of heat; willow, for example, will not produce a great deal of heat, whereas oak will produce a lot. The water content is also an obvious factor; if the fuel is too green or wet, a lot of its combustion energy will be used up converting the water to steam and not in increasing the chamber temperature. Creating too small an air intake in the firebox can also hinder climb. As has been mentioned in chapter 3, downdraft kiln fireboxes should have one-third space above the fire bars and two-thirds under; this should ensure that enough air can be drawn in to burn the fuel. Too much restriction of the chimney, or building the chimney too high or too short, can also compromise the effectiveness of a wood kiln.

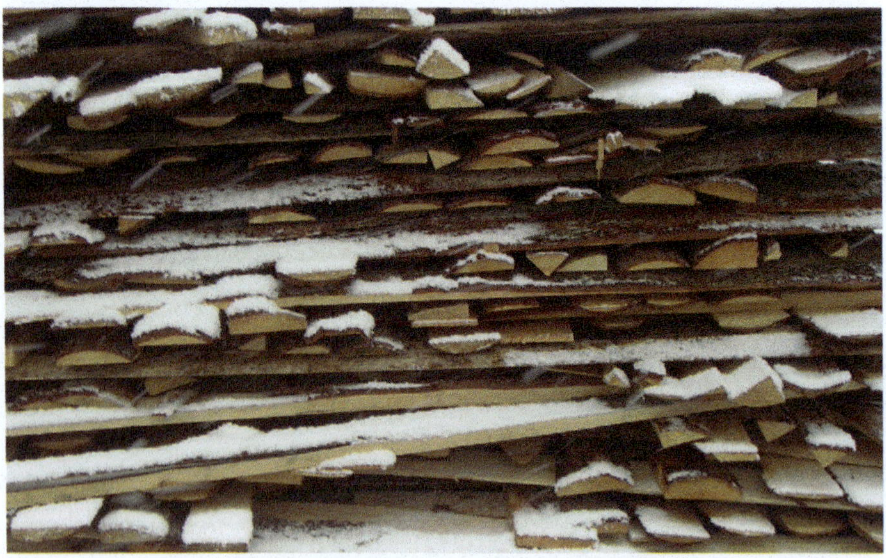

A cross-stacked wood pile seasoning. *Photo: Author*

Conclusion

The firing of clay and the structures which we use to carry this out could on the one hand be viewed as simply a technicality of the material which we choose to use. Hopefully what has been illustrated in this book is that on the other hand, the choice of kiln and firing technique can be as intrinsic to the creative making process as the clay itself, requiring as much consideration, concentration and engagement as the making itself.

Whether the kiln uses electricity, gas, wood or oil (or any other fuel for that matter) to generate heat, the way in which the maker engages with the attributes offered by that process makes their work individual to them. Hopefully this book will demystify some of the options, processes and techniques involved and will act as an inspiration to the possibilities available to us all as potters and ceramic artists.

Glossary

Aluminosilicate: Minerals which are made up of alumina, silica and oxygen making up a large part of clay materials.

Atomiser: The part of an oil-burner system which mixes forced air and oil and sprays the mixture as a fine mist.

Ball clay: Fine secondary clays.

Bentonite: Very plastic impure clay used for plasticising porcelain.

Biscuit ware: Work which has had a first firing and is no longer clay.

Ceramic change: An umbrella term which describes the chemical and physical changes as clay is fired and becomes ceramic.

Contactor: A switching device which delivers power to the elements of an electric kiln.

Crystobalite: A form of silica which develops in a clay body during firing. The higher the firing the more crystobalite is formed, excess crystobalite can make a body more prone to dunting.

Damper: A mechanical way of fully or partially closing the chimney on a fuel-burning kiln. Usually a kiln shelf, or brick.

Deflocculant: Added to a casting slip to increase fluidity.

Earthenware: A term which describes low-firing clay bodies or a glaze firing range between 1020°C and 1100°C (1868°F and 2012°F).

Energy regulator: Used on manual electric kilns to increase power to the elements.

Fire brick: Heavy, high-alumina fire brick. Used in kiln building.

Fly ash glaze: A glaze that occurs due to an accumulation of wood ash and flame borne alkalis when wood firing.

Green ware: Dry but unfired clay work.

HTI brick: Lightweight, high-temperature insulation brick.

Kanthal wire: Wire commonly used in the manufacture of kiln elements.

Leatherhard: Clay which has dried to the point that it can be handled without distortion, (firm and with the appearance of leather?).

Mouse hole: An aperture into a fuel-fired kiln; usually designed to introduce secondary air into the firebox.

Orton cones: A common brand of pyrometric cone.

Plastic clay: Clay in its soft, wet, malleable state.

Porcelain: A hard, white translucent clay body.

Props: Refractories used for supporting kiln shelves.

Pyrometric cone: A cone formulated to measure the heat work in a kiln firing.

Pyro-plasticity: Umbrella term that describes the softening of clay in high-temperature firings.

Quarl: Part of an oil-burner system. A refractory block with a conical aperture which ignites an oil and air mixture once heated.

Quartz: Silica.

Raku: A firing method whereby pots are removed from the kiln at orange heat and allowed to cool rapidly or immersed in combustibles.

Ramp: Describes the climb in temperature as part of an electric kiln-firing programme.

Refractory: Umbrella term which describes kiln shelves, props and other kiln furniture.

Refractory cement: Heat resistant cement used for repairing HTI brick kilns.

Saggar: A refractory vessel used for protecting work from kiln atmosphere or used to contain combustible materials to develop alternative surfaces.

Salt glaze: A glaze formed by the introduction of salt (sodium chloride), into a high-temperature kiln.

Salting port: Hole in a salt kiln positioned for the introduction of salt.

Setter: A supporting structure used to decrease distortion in ceramic wares.

Silica: Quartz/Flint.

Sintering: The fusing together of clay materials in a biscuit firing.

Spy hole: A hole positioned in a kiln to facilitate viewing of pyrometric cones or draw rings.

Stilts: Refractories designed to support glazed work in an earthenware glaze firing.

Stoneware: A term that describes a high-firing clay body or a glaze firing between 1220°C and 1400°C (2228°F and 2552°F).

Terra sigillata: A very fine slip with a natural sheen at low temperatures (900–1000°C/1652–1832°F).

Wadding: A soft mixture used to support work in salt firings and wood firings.

Kiln building materials suppliers

UK SUPPLIERS:

Fire brick, refractory castables and burners:

Kilnlinings
53, Ennerdale Close
Dronfield
Derbyshire
S18 8PL
Please note, the above address is for correspondence only
http://kilnlinings.co.uk/

Fire bricks, HTI bricks:

Refractory & Industrial Services Ltd
118, Longfellow Rd
Dudley
West Midlands
DY3 3EH
Tel: 01902 672995
Ris-ltd@btconnect.com

Refractory castables:

Resco
Resco Products (UK) Ltd
Newbold Works, Melbourne Rd
Lount
Ashby-De-La-Zouch
Leicestershire
LE65 1PL
Tel: 01530 222694
www.rescoproducts.com

General small kiln repairs:

Bath Potters' Supplies
Unit 18
Fourth Avenue
Westfield Trading Estate
Radstock, Nr Bath
BA3 4XE
Tel: 44 (0)1761 411077
Email: sales@bathpotters.co.uk

Gas burners:

Aeromatic-Barter Limited
Kynoch Road
Eley's Estate
Edmonton
London N18 3BH
http://www.cgsystems.co.uk/nlec/aeroma.htm

Oil burners:

Nu-way
Ten Acres
Berry Hill Industrial Estate
Droitwich
Worcestershire
WR9 9BP
Tel: +44 (0) 1905 794 331
Email: info@nu-way.co.uk
http://www.nu-way.co.uk/burners.asp

US SUPPLIERS:

Refractory castables:

Resco
2, Penn Center West; Suite 430
Pittsburgh
PA 15276
888-283-5505
www.rescoproducts.com

Gas burners:

Ward Burner Systems
PO Box 1086
Dandridge,
TN 37725
(865) 397-2914 phone
(865) 397-1253 fax
info@wardburner.com
http://www.wardburner.com/

Refractory castables:

Allied Mineral Products, Inc
2700 Scioto Parkway
Columbus
Ohio 43221
http://www.alliedmineral.com/products.html

Gas and oil burners:

Wayne Combustion Systems
Home Office Address:
801 Glasgow Avenue
Fort Wayne
IN 46803
Email: sales@waynecs.com
http://www.waynecombustion.com/default.aspx

Fire bricks and castables:

Hi Temp Refractories Co
2250 N Lark Drive
Fenton
MO 63026-4329
Phone: 636-677-2200
Email: sales@hitempincusa.com
http://www.hitempincusa.com/index.asp

Orton pyrometric cone temperatures

Large regular cones

Pyrometric cones measure heat work and so depend on the heating rate. A slower temperature rise will cause the cone to bend at a lower temperature.

Cone no.	60°C/hour	108°F/hour	150°C/hour	270°F/hour
09	917	1683	928	1702
08	942	1728	954	1749
07	973	1783	985	1805
06	995	1823	1011	1852
05	1030	1886	1046	1915
04	1060	1940	1070	1958
03	1086	1987	1101	2014
02	1101	2014	1120	2048
01	1117	2043	1137	2079
1	1136	2077	1154	2109
2	1142	2088	1162	2124
3	1152	2106	1168	2134
4	1160	2120	1181	2158
5	1184	2163	1205	2201
6	1220	2228	1241	2266
7	1237	2259	1255	2291
8	1247	2277	1269	2316
9	1257	2295	1278	2332
10	1282	2340	1303	2377
11	1293	2359	1312	2394
12	1304	2379	1324	2415
13	1321	2410	1346	2455
14	1388	2530	1366	2491

Further reading

Fraser, H. *Electric Kilns and Firing*, Pitman, 1979

Gregory, I. *Kiln Building*, A&C Black, 2006

Gregory, I. *Alternative Kilns*, A&C Black, 2005

Hessenberg, K. *Sawdust Firing*, University of Pennsylvania Press, 1994

Jones, D. *Firing: Philosophies within Contemporary Practice*, Crowood Press, 2007

Minogue, C. and Sanderson, R. *The Log Book*, http://homepage.eircom.net/~thelogbook/

Olsen, F. *The Kiln Book*, Chilton, 1983

Rogers, P. *Salt Glazing (Ceramics)*, A&C Black, 2002

Watkins, J. and Wandless, P. *Alternative Kilns and Firing Techniques*, Lark Ceramics Books, 2006

Index

Printed and bound by CPI Group (UK) Ltd, Croydon, CR0 4YY
14/09/2026
02216526-0001